chambre syndicale de la
recherche et de la production
du pétrole et du gaz naturel

comité des techniciens

commission exploration
sous-commission contrôle des sondages

GEOLOGICAL AND MUD LOGGING IN DRILLING CONTROL

catalogue of typical cases

1982

ÉDITIONS TECHNIP 27 RUE GINOUX 75737 PARIS CEDEX 15 technip

ISBN 2-7108-0425-5

Printed in France
by Imprimerie Hemmerlé, 75002 Paris

The English version was jointly sponsored by the Institut Français du Pétrole-Ecole Nationale Supérieure du Pétrole et des Moteurs and by Geoservices.

The English translation was prepared by J.P. Heuzé assisted by J. Hamilton.

INTRODUCTION

This "Catalogue of Typical Cases" was written between 1977 and 1981, based on examples supplied by the Compagnie Française des Pétroles (TOTAL-CFP) and Société Nationale Elf Aquitaine-Production (SNEA-P).

Experts from the petroleum industry prepared the studies within the framework of the "Geological Drilling Control" sub-committee which is attached to the "Engineers Committee" of the Chambre Syndicale de la Recherche et de la Production du Pétrole et du Gaz Naturel. The sub-committee was headed successively by A. Pochitaloff (SNEA-P) and M. Verdier (Institut Français du Pétrole - Ecole Nationale Supérieure du Pétrole et des Moteurs).

Since 1945, the Engineers Committee has conducted studies and issued recommendations that have led to standardized practices in the various fields of the petroleum industry: exploration, development, environment, safety, documentation, etc.

The results of the committee's work are disseminated in the form of manuals, articles in professional magazines and textbooks published in the ENSPM collection.

TABLE OF CONTENTS

CASE 1

BEGINNING OF AN OIL KICK

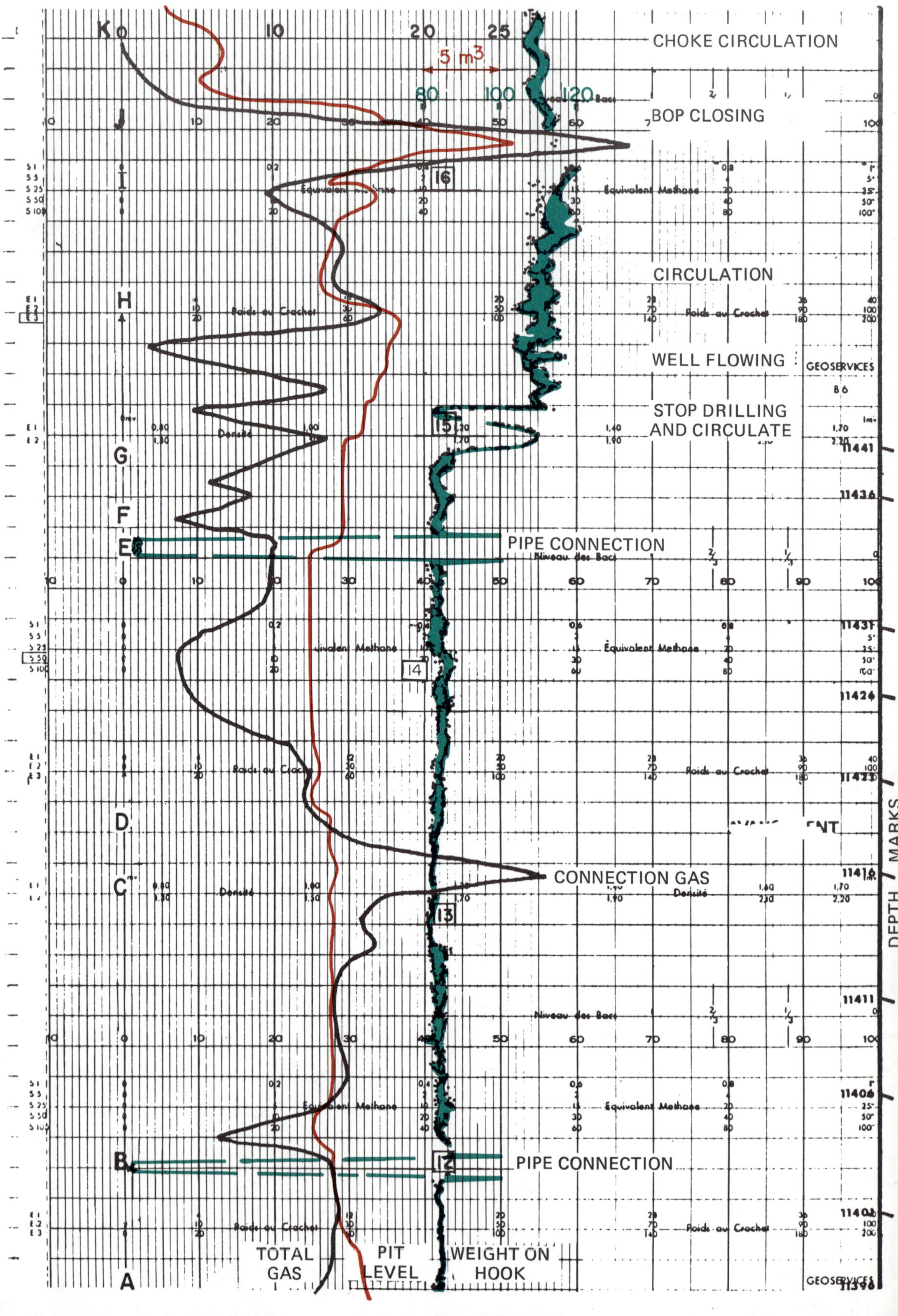

LOGS SHOWING THE BEGINNING OF AN OIL KICK

This example shows the "multi-track" recording of the interval between 11 396 and 11 441 feet of a well drilled in 1972 in a dominantly shaly deltaic series with thin sandy and dolomitic crossbedding.

1. TECHNICAL DATA

1.1. Drilling

- 9 5/8" casing shoe at 10 655'
- 8 1/2" drill bit; mud density d = 1.42 to 1.56
  Note: The use of 6 1/4" drill collars in a 8 1/2" hole increases the danger of swabbing a possibly producing zone.
- Mud flow while drilling: 1 600 l/min for an average annulus volume of 30 l/m, giving a theoretical lag time of about 60 minutes.

1.2. Logging

Drilling control is ensured with the following equipment:

- Rate-of-penetration recorder with depth display
- Degasser and conductivity total gas detector
- Chromatograph for gas analysis
- Weight-on-bit recorder
- Mud-pit level recorder

1.3. Recorded parameters and scales

The "multi-track" log is recorded on a 280-millimetre-wide chart unwinding vertically from top to bottom versus time; the space between two horizontal lines corresponds to a period of 7 min 30 sec.

- The depth marks are recorded every 5 feet in the right-hand margin.

- The total gas curve is recorded on the S 50 scale, corresponding at full scale to 50% equivalent methane in the analyzed gas-air mixture. On that scale, one large division corresponds to 5% and one small division to 0.5%.

- Mud-pit level variation is recorded on a measurable 50 m^3 full-scale range; one large division corresponds to 5 m^3 and a small one to 0.5 m^3.

- The weight on hook is recorded on a 0-200-ton scale; a large division corresponds to 20 t, a small division to 2 t.

2. OPERATION DISCUSSION

A-B 11:30 to 12:00 hrs

Drilling at an average rate of penetration of 4 min/ft. Between 11:45 hrs and 11:50 hrs, mud loss of 3-4 divisions corresponding to 1.5 to 2 m^3 while drilling through a thin porous bed.

B-C 12:00 to 13:07 hrs

From 12:00 hrs to 12:05 hrs, pipe connection at 11 403'. The weight on hook increases sharply coming off-bottom and then decreases as the drill-string is laid on slips.

This pipe connection is also clearly indicated on the total gas curve, but with a 4-min lag time corresponding to the transit time of the gas from the degasser to the detector.

A slight variation in mud-pit level is also noticeable at pipe connection.

From 12:05 hrs to 13:07 hrs, drilling from 11 403' to 11 415' at 4-5 min/ft. Scattering of the weight-on-hook curve is due to readjustment of the weight on the bit by the driller as drilling proceeds at faster rates.

C-D 13:07 to 13:20 hrs

Arrival at the surface of a gas slug accumulated while connecting the pipe at 11 403'. The shift is in accordance with the theoretical lag time.

D-E 13:20 to 14:30 hrs

Drilling from 11 418' to 11 434' with a slight mud loss (1 m^3) while drilling through a thin porous bed at 11 420'. The steady mud loss (0.5 m^3) is due to the filling of the section of hole drilled during that time interval, and the increase in gas percentage between 14:15 and 14:30 hrs is most probably due to the recycling of the gas slug formed at the 11 403' pipe connection and observed at 13:07 hrs.

E-F **14:30 to 14:35 hrs**

Pipe connection at 11 434'. The weight-on-hook and total gas curves react as previously. On the other hand, the mud-pit level curve shows a sharp increase of 5 divisions (2.5 m^3) in 2 minutes (flow most probably started by swabbing when the drill-string was brought off-bottom for pipe connection).

F-G 14:35 to 14:55 hrs

Drilling from 11 434' to 11 441'. The drilling rate increases 50%. The mud- pit level remains stable; this means that the formation is still producing slightly while drilling, because the increase in hole volume is not evidenced by a decrease in the mud-pit level as the hole is being filled.

G-H 14:55 to 15:30 hrs

In view of this slight inflow, drilling and circulation are stopped to observe the well. Taking the bit off-bottom gives an instant mud gain of 1 m^3 (swabbing at pick-up).

The well flows for half an hour. A steady increase of the mud-pit level is observed (total gain: 6.5 m^3 since the first gain). On the other hand, the gas curve shows apparently erratic gas peaks; in fact they do correlate perfectly with the various drill-string displacements.

H-I 15:30 to 16:00 hrs

When circulation is resumed, the mud-pit level falls abruptly, and after 5 or 6 minutes the well starts flowing again, slowly at first, and then more rapidly (4 m^3 gain in 20 minutes and then 12 m^3 in 7 minutes).

I-J 16:00 to 16:15 hrs

The 34% equivalent methane gas peak associated with the significant mud gain corresponds to the arrival at the surface of the oil and gas produced by the formation at 14:30 hrs while connecting the pipe at 11 434'.

The gas show delay is correct: it agrees with the lag time (60 min) added to the time the circulation was stopped (30 min).

J-K 16:15 to 16:37 hrs

At 16:15 hours the decision is taken to close the blowout preventers and to circulate under choke to evacuate the gas and to weight the mud. From that time, the gas detector is not fed anymore and the total gas curve goes to zero (correct return to zero without any drift).

3. RECORDER FUNCTIONING

The overall quality of the logs is good. The following points are to be noted:

- The transit time of the gas (from degasser to detector), even if added to the dilution time, seems a bit long (4 minutes).

- No zero check of the gas curve was made during these 5 hours of recording. However, the return to zero is correct at the end of the recording.

- The scattering of the points on the weight-on-hook curve reflects the weight-on-bit adjustments made by the driller with the brake each time the weight diminishes as drilling progresses.
  Actually, to record the weight-on-hook curve, a reading is taken every 20 seconds, which means that the measurement is not always made in the same condition: there is either some weight loss (due for example to the bit penetration), or some weight gain (due for example to the driller's weight adjustment). This is mostly true for soft formations. For hard formations, the scattering of the weight-on-hook measurements may have another explanation: it corresponds to the jarring motion of the kelly due to the jolting of the bit as it bites into the formation.

4. COMMENTS

4.1. Drilling operation discussion

In spite of the proper functioning of the logging equipment and correct interpretation of the recorded parameters, the well was lost: the fluid inflow caused the unconsolidated formation sloughing, and sticking of the drill string.

The loss of the well could most probably have been prevented through strict compliance with the instructions given to the drillers:

- Shutting-in of the well on first occurrence of mud gain at 11 433' to measure the shut-in annulus pressure to identify the flowing fluid (gas or liquid).

  With the well shut-in and circulation stopped, if the fluid is a liquid, the shut-in casinghead pressure will be constant if there is no gas at all. If the bit is at T.D., the shut-in drill-pipe pressure added to the hydrostatic pressure of the mud column in the drill pipes gives the formation pressure.

  On the other hand, if the fluid produced by the formation is gas, the casinghead pressure increases until all the gas has reached the surface, except of course, in case of fracturation of the formation and/or casing.

- Circulation at 11 4334' on first indication of surface flow, or at least choke circulation to weight the mud when the second gain was noticed at 11 441'.

  This should have restricted the inflow of fluid and would most probably have prevented the loss of the well.

4.2. Type of fluid

The linear and steady increase of the mud volume at the surface suggests that the fluid is a liquid (oil or water) and not a gas. Had the formation been producing gas, the mud gain at the surface would have been much larger and accelerating because of the gas-expansion mode in the annulus.

DETECTION OF A WATER KICK AFTER SWABBING AT A PIPE CONNECTION

1. TECHNICAL DATA

1.1. Drilling

- 13 3/8" casing shoe at 3 165 metres
- 12 1/4" bit
- XB23 biopolymer mud; density 1.01 to 1.03; V = 35; F = 10; ClNa = 0.5 g/l
- Mud-flow rate 2 250 l/min; approx. lag time 75 minutes
- Weight-on-bit 14 to 18 tons; rotation 60 to 70 rpm (with 12 1/4" square stabilizer)

1.2. Type of formation

Carbonate series: grey microsparite, slightly shaly. Cutting analysis shows a large amount of calcite (5 to 20%), which may suggest a formation fracturation. From 3 240 to 3 250 metres, the formation is more shaly and made of a compacted shaly micrite, dark grey to black.

1.3. Logging

The following logs were used in this example:

- Mud-pit level, with one float-sensor per pit
- Automatic recording of the penetration
- Chromatograph for gas analysis

1.4. Recorded data and scales

- In the right-hand margin, depth is shown for each drilled metre
- On the right-hand half of the chart, the mud-pit level is shown with a 1 m^3 per-division scale
- On the left-hand half of the chart, the chromatographic recording is shown with an indication of the scale for the methane contained in the analyzed mixture; the envelope of the C_1 peaks is drawn.
- The time is indicated along the centre line of the chart; one horizontal division equals 15 minutes.

2. OPERATION DISCUSSION

Drilling was recently resumed after setting the casing: less than 100 metres of open-hole. Mud conditioning is not completed. 13 m^3 of mud was added between 3 242.80 m and 3 243.80 m. Down to 3 252.40 m, the mud-pit level curve shows slight variations with a wavy aspect and a tendency toward losses partly masked by the mud processing.

Drilling progresses at an average speed of 3 to 4 metres per hour.

A-B **23:30 to 23:40 hrs**

Stopped drilling at 3 252.40 m for pipe connection. An instantaneous mud gain of 1.2 m^3 is noticed at 23:30 hrs.

B-C 23:40 to 0:35 hrs

Drilling is resumed at 23:40 hrs. Mud gain requires a zero shift of the mud-pit level curve at 0:10 hrs.

To check the well behaviour, drilling is stopped at 3 255.20 m, after a 4.5 m^3 gain since drilling was resumed after the last pipe connection.

C-D **0:35 to 1:50 hrs**

The mud is kept circulating under normal conditions, and an arrival of C_1 shows on the chromatograph curve at 1:00 hrs, i.e. 80 minutes after pipe connection (a maximum reading of 1.5% C_1 is recorded while the gas is being transferred). A first transfer of 5.6 m^3 is done under normal circulation conditions.

D-E 1:50 to 2:20 hrs

Choke circulation. During this operation (started at 1:50 hrs), 11.6 m^3 of contaminated mud is transferred. The contaminating fluid is analysed and identified as being salt water with 333 grams per litre ClNa and 9 grams per litre Ca, at a temperature of 55°C.

The drill-pipe shut-in pressure measurement indicates 20 bars (300 psi).

DETECTION OF A SALT-SATURATED WATER KICK

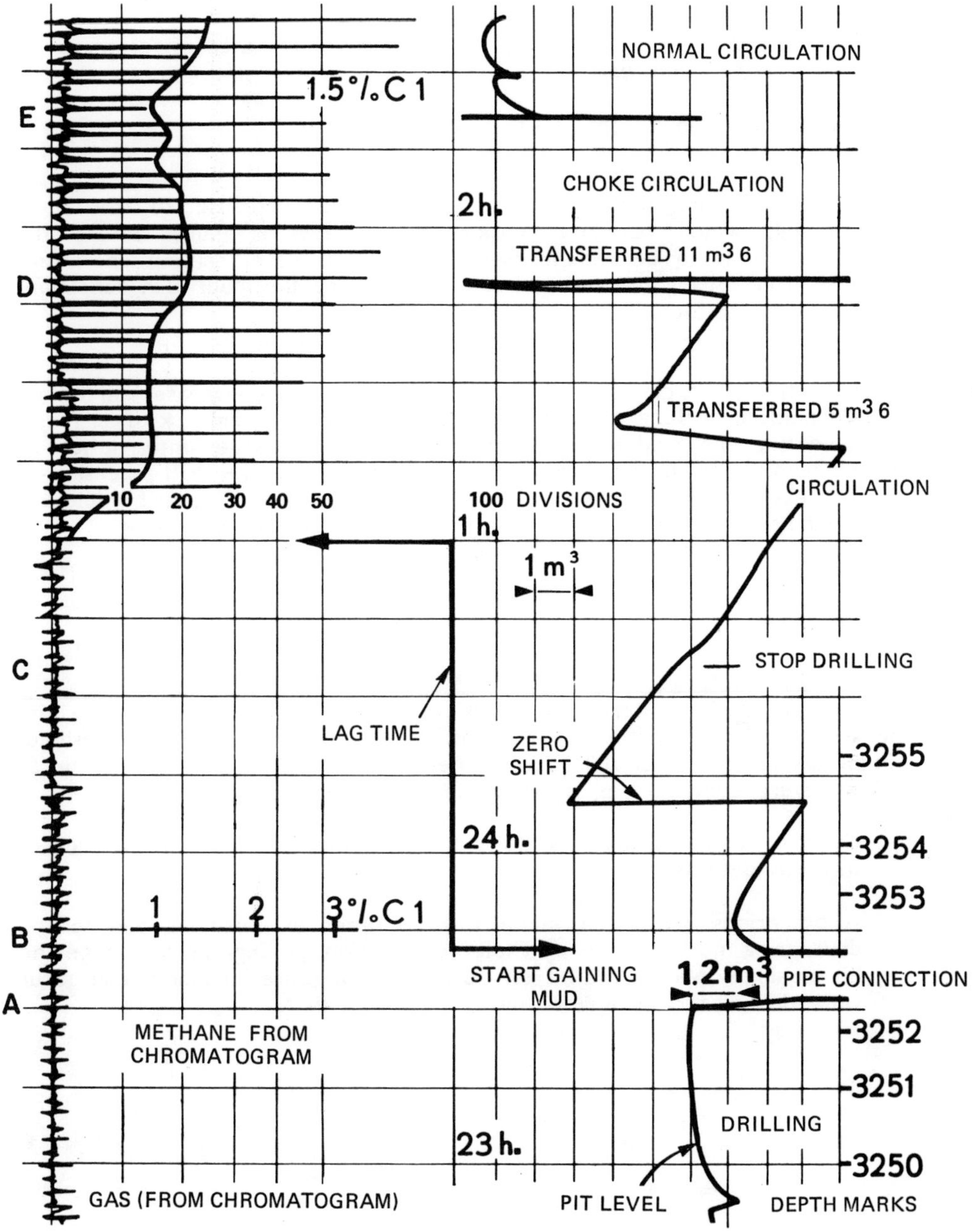

3. COMMENTS

a) Pore pressure calculation

At 3 250 metres, a mud with a density of 1.03 gives an hydrostatic pressure of 334.7 bars. The pore pressure is thus 334.7 + 20 = 354.7 bars.

A downhole pressure test of this section indicated a pressure of 354.69 bars at 3 247 metres.

Weighting the mud from 1.03 to 1.10 is a must.

b) Monitoring the mud volumes was difficult because of the mud conditioning, which partially masked the losses or gains before the 1.3 m^3 gain. It is to be remembered that in this case, drilling operations resumed after setting the casing and reconditioning the mud; such conditions require close attention.

c) In spite of the clear indication of the mud-volume increase starting at pipe connection, drilling went on for almost one hour, at the risk of seriously contaminating the mud with an unknown fluid. An earlier interruption of drilling to monitor the mud volume at the surface and the drill-pipe shut-in pressure evolution would have indicated the type of fluid. The test would not have been necessary knowing the fluid type and its calculated pressure.

4. CONCLUSION

Mud-pit level monitoring is a very useful log. Modern equipment permits the detection of very small variations. However, the detection threshold, which may be as low as 100 litres in the case of an on-shore rig, may be as high as several cubic metres for an offshore drilling rig:

- Gains or losses, always due to a pressure imbalance, are in most cases instantly detected.
- A precarious pressure balance will be upset by swabbing due to the drill-string motion in the well (pipe connection, pulling-out to change the bit or simply pulling the bit off-bottom). A pressure imbalance can occur while drilling through a reservoir or even after that.

USING THE CHROMATOGRAPH TO DETECT ABNORMAL PRESSURES IN UNDERCOMPACTED SERIES

The difficult drilling conditions usually encountered in undercompacted shaly formations (over-pull, rimming, gas-cut mud, etc.) are often associated with a gradual change in the composition of the gas contained in the mud, evidenced by decreasing ratios C_1/C_2, C_1/C_{2+}, C_2/C_3, etc.

One often observes C_2/C_3 lower than or equal to one.

This kind of gradual change or evolution, which can generally be associated with some sort of drilling difficulty, is often a precursor of more serious problems.

The two examples that follow illustrate such an evolution. To a certain degree, this evolution is tied to the Δp applied to the formation which may also change the quality and/or the representativeness of the gas indexes. However, an increase in maturation of the organic matter, also translated by such an evolution, could be associated with a significant increase in pore pressure (abnormal pressure) in thick shaly beds.

EXAMPLE 1 (Fig. 1)

In this example, the change in the composition which appears at 1 800 metres, i.e. about 100 metres before the occurrence of drilling problems (over-pull, reaming, gas-cut mud) may indicate that an over-pressured zone is being drilled through.

The tests conducted in this well confirmed a change in pressure gradient.

Increasing the mud density, and thus the Δp applied to the formation, eliminated the drilling problems and restored the initial composition.

EXAMPLE 2 (Fig. 2)

1) C_2/C_3 curve

This curve was drawn from the chromatograph readings taken every 10 metres in the gassy zone. The gas indexes are only those of the shales; the values corresponding to the sand and lignite layers were not used (only indexes from identically permeable levels may be validly compared).

At G1 (3 000 metres), the curve shows a sharp decrease in the ratio from 1.7 to 0.9, to 0.7, and then to 0.5. It then stays between 0.5 and 0.6.

At G2 (3 360 metres), the ratio falls to 0.2 at the beginning of the kick, before choke circulation. It goes back to values higher than 2.0 as soon as the mud density is increased.

2) Output mud temperature

Each segment gives the profile of the recording made during a bit run.

3) Temperature gradient

The curve is drawn using the output temperature values. The calculation may be made at regular intervals (e.g. every 10 m of penetration) or for each bit run. This last method was used here, as there were many bit changes.

The gradient is given in degrees Celsius for 100 metres:

$$G°C/100 = (t_2 - t_1)\ 100\ /\ h$$

- t_1 and t_2 are the temperatures at the beginning and at the end of the bit run, in degrees Celsius,

- h is the drilled interval in metres.

4) Pressure profile taken from the tests

The points used to plot this profile are the values measured while testing the well at various depths. The extrapolation of the curve between points 4 and 5 is not realistic: it is most probably closer to the line shown with question marks.

5) Mud pressure

This is the recording of the hydrostatic pressure of the mud versus depth.

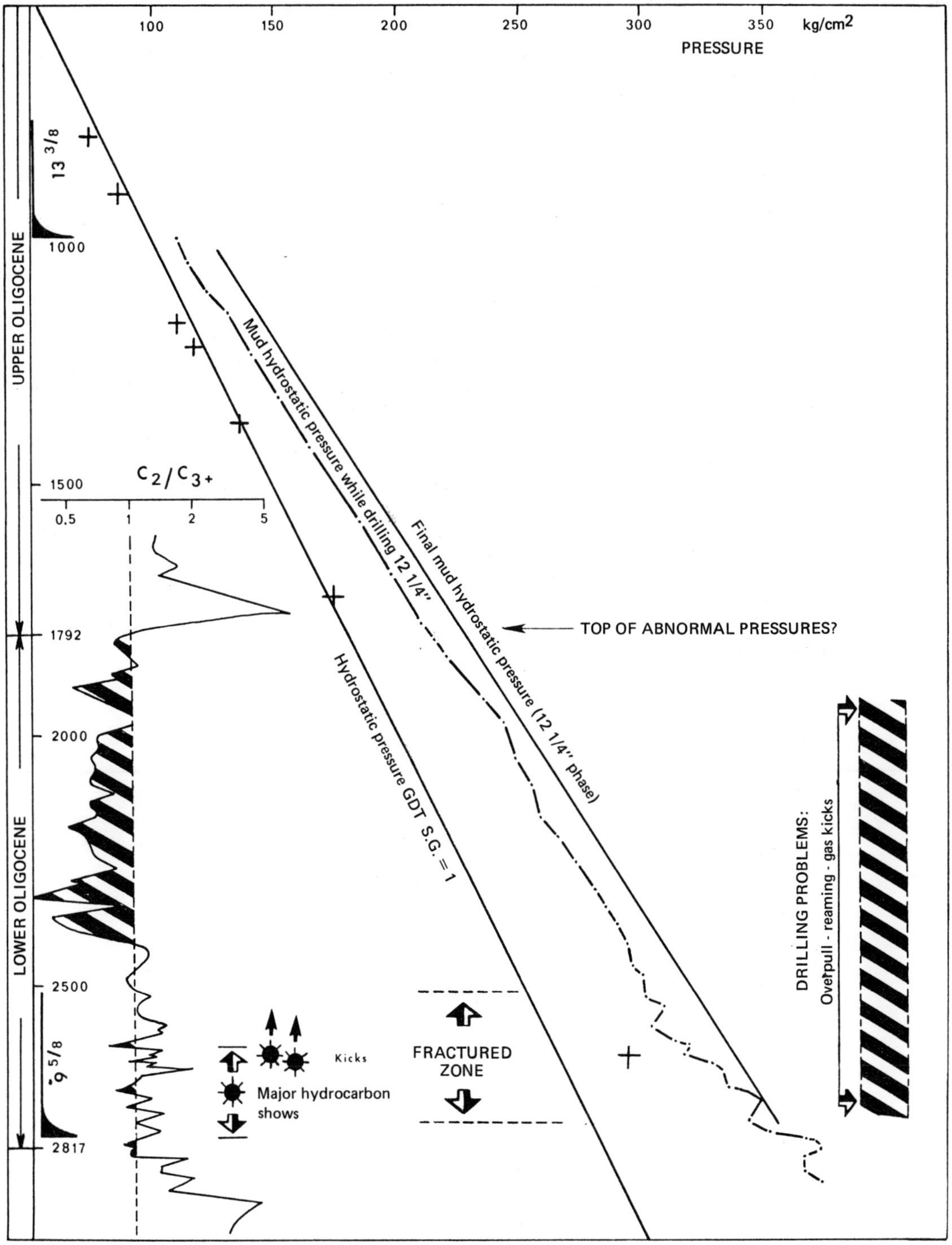

+ **Pressure test.**

Fig. 1.

6) Test temperature

As for the pressure profile, the points where taken from the test values. They show the same trend as those of profile 7. Notice the slope change at the top of the abnormal pressures.

7) Corrected temperature (T_C)

The corrected temperature was calculated from the various gradient values obtained previously (curve 3). The formula is:

$$T_C \,^{\circ}C = (gradient/100 \,.\, h) + T_{corr}$$

T_{corr} is the corrected temperature at the end of the previous bit run. Depending on the situation, the surface temperature may be the average surface temperature or the sea bottom temperature (offshore). It is often convenient to take the temperature recorded during the electrical logging operations, before the last casing section is set. A simpler solution consists in end-to-end plotting of the various segments of curve 2: this is the "end-to-end" curve produced by most geological control service companies.

The more numerous the bit changes (or any other drilling interruptions), the more disturbed the corrected temperature curve will be. In such cases, the corrected temperature might be much higher than the formation temperature. This is what happened in this example. In fact, one does not use this curve to obtain the formation temperature, but only its profile, or at least that of the hole-bottom temperature.

COMMENTS

- The first increase in the temperature gradient (T1) could not be satisfactorily explained, except as a change in the deposition medium.
- The second increase in the temperature gradient (T2) gave a warning 60 metres above the danger zone.
- As the gas-composition ratio started decreasing (G1), an increase in the formation pressure gradient should have been suspected (transition zone), about 350 metres above the top of the eruptive zone.

CASE 3

PRESSURE-TEMPERATURE PROFILES

Fig. 2.

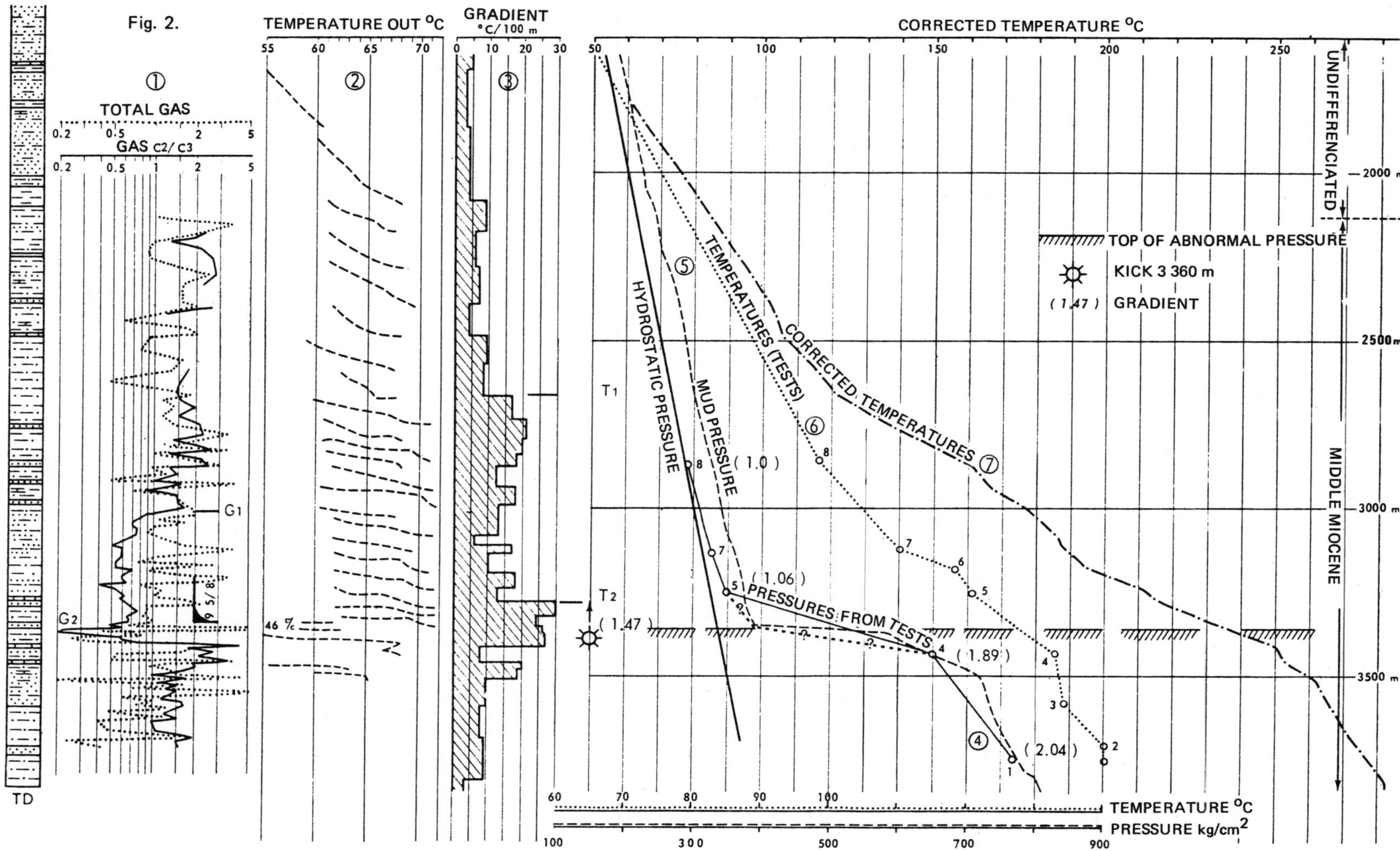

USING CALCIMETERS

CORRELATION WHILE DRILLING

The calcimetry curve has always been an excellent correlation tool in predominantly carbonaceous formations. Unfortunately, this log is not always available for the well with which one tries to get a correlation.

USING THE INTEGRATED SONIC

It is out of the question to directly compare the calcimetry curve with the much more detailed "Δt" sonic transit time curve.

On the other hand, excellent results may be obtained using the transit time integrated over depth increments (e.g. 10 metres).

Such a curve may be easily and quickly plotted using the integration peaks of the sonic log.

Plotting the curve on a reduced scale, say 1/10 000, gives a better contrast and reveals certain characteristics that are difficult to notice on a 1/500 scale (such as the slow decrease in carbonate content between the α and β marks in the illustration).

EXAMPLE OF USE

The three wells in this example were drilled in a formation essentially made of shales and carbonates, with a very monotonous lithology.

CALCIMETER-SONIC CORRELATION

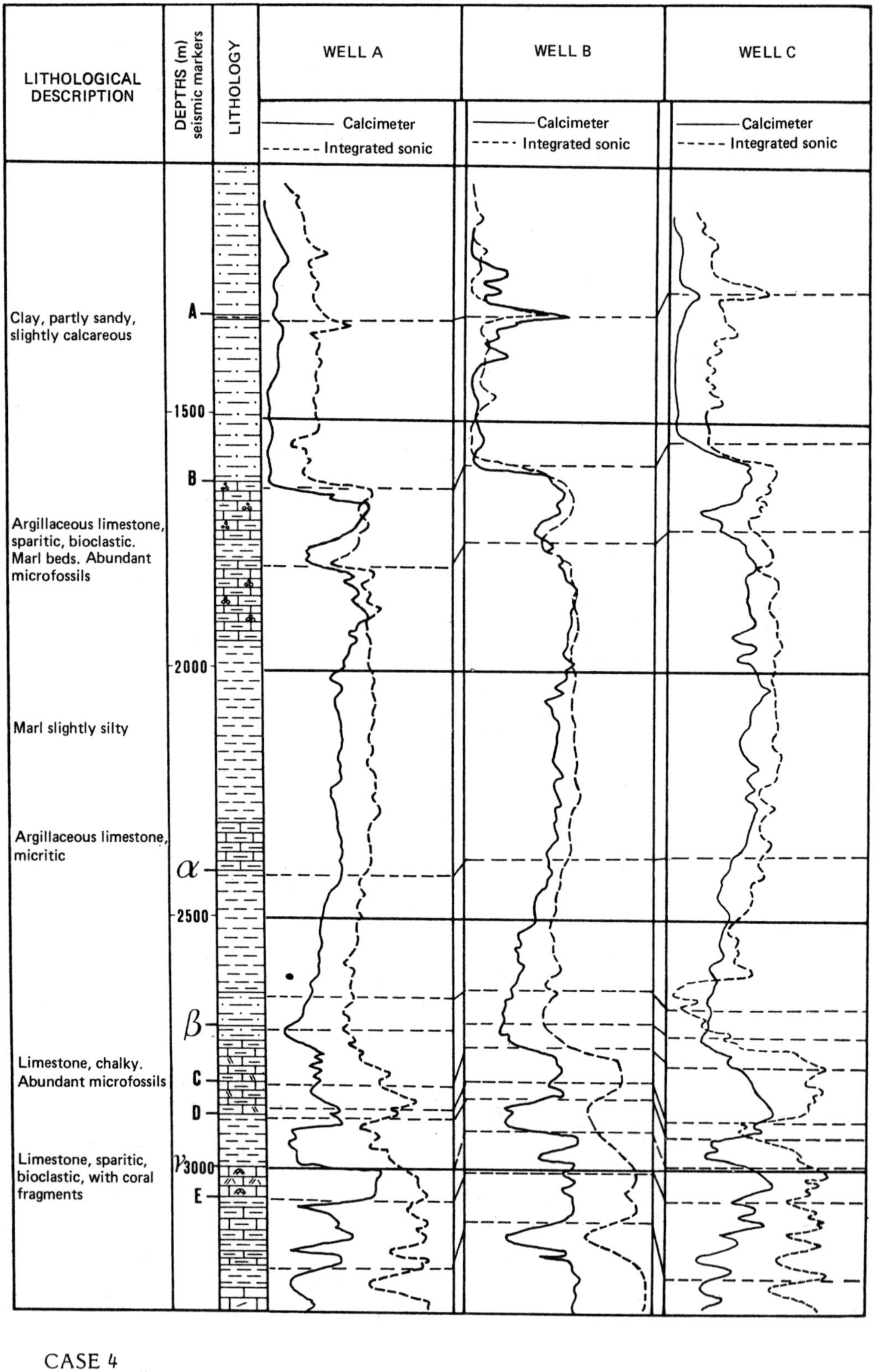

CASE 4

PLOTTED CURVES

- Calcimetry: on the left (plain curve).
- Integration: on the right (dotted curve).
- The limits and correlations indicated are mostly based on the micro-paleontological analysis.
- Levels A, B, C, D and E are seismic markers.

Notice that the integration curve is "driven" by the carbonate contents and that there is no difficulty in correlating the calcimetry curve of the well being drilled with the sonic integration curve of another well drilled in the area:

- The calcimetry curve on its own allows the geologist to know "where he is" in comparison with other wells, and to identify the main seismic markers "as they come".

- The top of the zone of interest is indicated here by the "D" marker: it is clear that the calcimetry and integration curves are sufficiently contrasted and correlatable to warn the geologist in due time.

ENTRY INTO A GAS RESERVOIR AND KICK CONTROL WHILE TRIPPING

1. TECHNICAL DATA

1.1. Drilling

This example shows an offshore well drilled in 1975 at a water depth of 300 metres. The 9 5/8" casing shoe is at 2 360 metres. Drilling is proceeding with a 8 1/2" bit and an input-mud density of 1.19. The pump flow-rate is 1 800 l/min. The mud return time is 45 minutes.

1.2. Logging

The multitrack recording was made on a 280-millimetre-wide chart unwinding vertically from top to bottom as a function of time. The vertical scale is 7 min 30 s for 1 centimetre (between two horizontal lines).

1.3. Recorded data and scales

The curves shown on the diagram are:

- Total gas (dotted line). The log is first recorded on the "S5" scale, corresponding to 5% full-scale methane equivalent in the analyzed gas-air mixture.

  At 06:00 hrs the scale is changed to "S25," corresponding to 25% full-scale methane equivalent in the analyzed gas-air mixture.

- The output mud density (broken line). The scale goes from 0.75 to 1.75. One large division equals 0.10.

- The mud-pit level (plain line). This gives the total of pit 1 + 2 + 3. One large division corresponds to 1 000 litres, i.e. 200 litres per small division.

- The pump on/off condition is shown on the right-hand side of the chart. The line is interrupted when the pumps are stopped.

- The depth marks (2 marks per metre) are shown in the right-hand margin.

2. OPERATION DISCUSSION

A-B 05:37 to 06:00 hrs

Drilling after pipe connection at 2 666 metres. The variation in the mud-pit level resulting from the pipe connection is noticeable. As drilling is resumed, the level shows a steady increase in volume. Drilling was stopped because of this increase and also because of a high drilling speed. Circulation is maintained.

B-C 06:00 to 07:19 hrs: circulating

At 06:10 hrs, the pit level sharply decreases and then increases again, suggesting the transfer of a gas slug.

At 06:22 hrs, the total-gas curve sharply increases. A scale change at 06:25 hrs does not allow reading of the maximum value of the peak B1. The density slightly decreases from 1.15 to 1.13.

These changes indicate the arrival of gas at the surface. The pit level decrease recorded 15 minutes before B1 agrees with the arrival in the riser and corresponds to the decrease in the gas volume due to the temperature change. Temperature measurements at 2 600 metres indicate 66°C (electrical logging) and 80°C (test). Sea temperature is 3°C; output mud temperature is 9°C.

The travel time in the 16" riser may be calculated:

$$\frac{\text{riser annulus volume}}{\text{pump flow rate}} = \frac{100\ \text{l} \times 300\ \text{m}}{1\,800\ \text{l/min}} = 16\ \text{min}$$

corresponding to the delay previously observed.

At 06:38, stopping the circulation gives a 700-litre increase in mud-pit volume.

From 06:45 till 06:50 hrs, arrival of a second gas slug reaching 16.5% methane equivalent. The density decreases to 1.09. Afterwards the pit level stabilizes and even shows a slight decrease.

At 07:07 hrs the degasser is cleaned, causing the gas curve to go to zero.

C-D 07:19 to 08:15 hrs

At 07:19 stopped circulating and started pulling out. The pit volume increase is due to various causes: complete emptying of the mud-return line, stopping of the shale shakers, filling of the possom-belly tank. This increase is followed by a decrease corresponding to the beginning of the trip out and filling of the hole. The density curve falls to zero as the feed-tank empties.

From 07:47 to 07:48, a gain of 1 m^3 is recorded while pulling out.

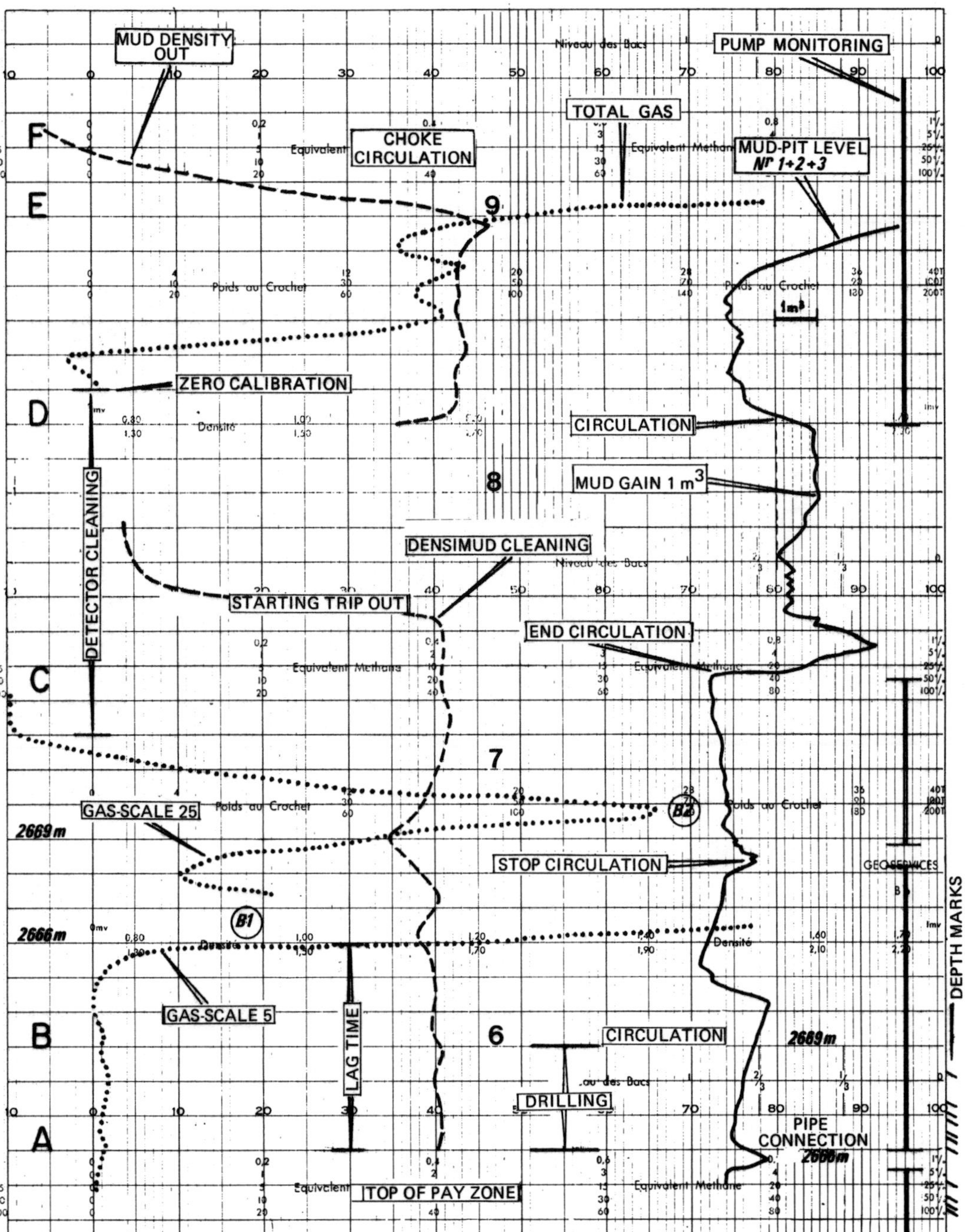
MUD DENSITY OUT
PUMP MONITORING
Niveau des Bacs
TOTAL GAS
CHOKE CIRCULATION
MUD-PIT LEVEL Nr 1+2+3
Equivalent Methane
Poids au Crochet
1m³
ZERO CALIBRATION
Densité
CIRCULATION
MUD GAIN 1 m³
DETECTOR CLEANING
DENSIMUD CLEANING
STARTING TRIP OUT
END CIRCULATION
GAS-SCALE 25
2669m
B2
STOP CIRCULATION
GEOSERVICES
DEPTH MARKS
B1
2666m
GAS-SCALE 5
LAG TIME
CIRCULATION
DRILLING
PIPE CONNECTION
2668m
TOP OF PAY ZONE
A
B
C
D
E
F
6
7
8
9

D-E 08:15 to 09:00 hrs

At 08:15, stopped pulling out and started circulating to control a possible kick. This led to a decrease in pit volume which is then stable until 08:40 hrs. The densimeter and degasser are put back into service.

Starting at 08:40 hrs, the mud-pit volume increases considerably and quickly (4 m^3 in 15 minutes), indicating the arrival of gas.

E-F 09:00 hrs

Arrival of the gas slug at the surface. The total gas curve suddenly increases and the output density decreases. Maximum values are not reached as choke circulation is started, which cuts the mud supply to the logging equipment. At the moment of going to choke circulation, the total gain was 22 m^3.

3. COMMENTS

This example shows that, even during a trip, the mud-logging personnel should keep a close watch. In this particular case, the volume increase was slight and went unnoticed by the driller who was busy with the pulling-out operation.

It is interesting to note that the arrival of the gas in the riser which is cooled by the sea is indicated by a decrease in pit level.

In this example, the p applied to the formation was only 10 bars and contributed to the swabbing process at the beginning of the trip.

CASE 6

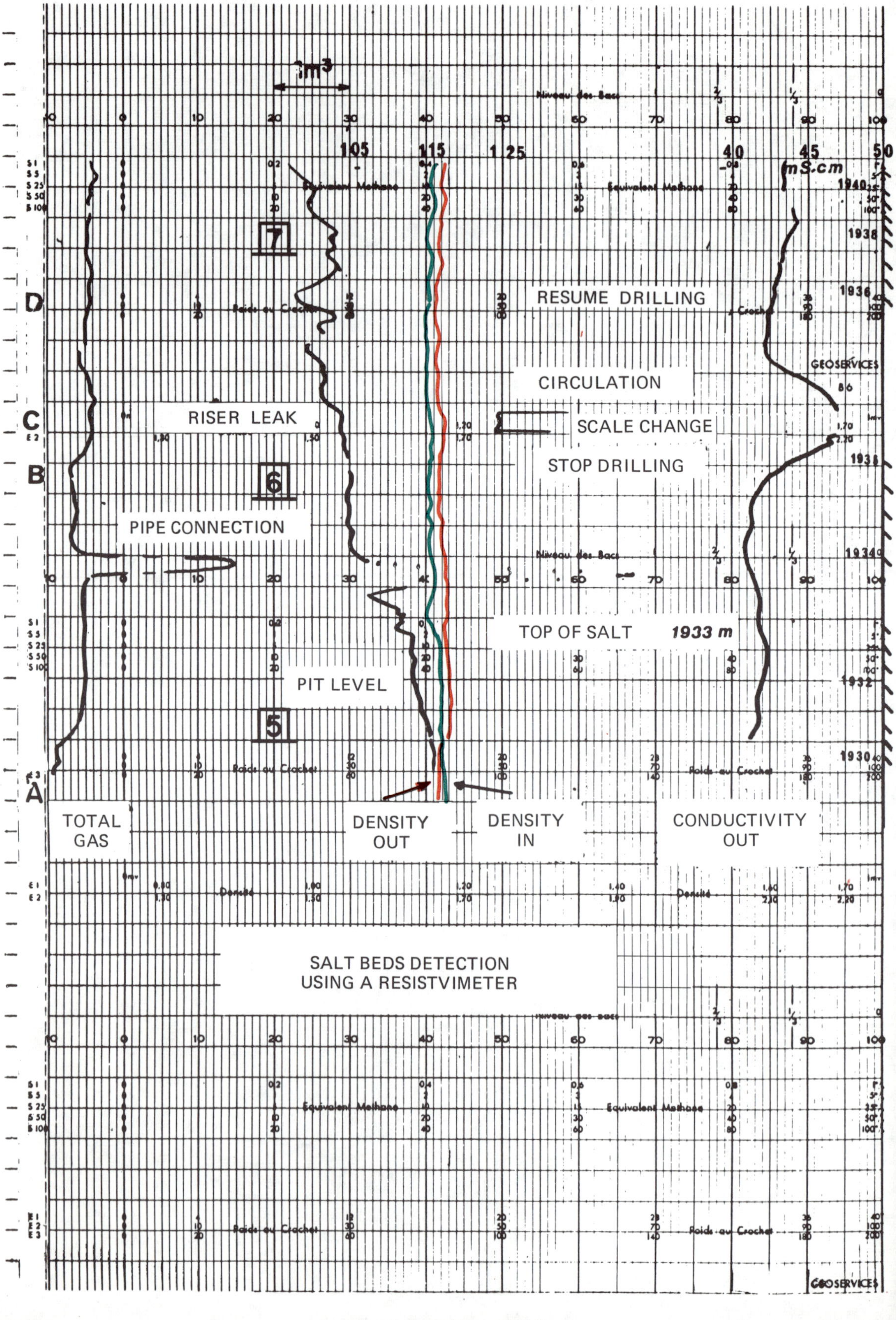

m³
105
115
125
40
45
50
mS.cm
1940
1938
1936
1934
1932
1930
7
6
5
D
C
B
A
RESUME DRILLING
CIRCULATION
RISER LEAK
SCALE CHANGE
STOP DRILLING
PIPE CONNECTION
TOP OF SALT
1933 m
PIT LEVEL
TOTAL GAS
DENSITY OUT
DENSITY IN
CONDUCTIVITY OUT
SALT BEDS DETECTION USING A RESISTVIMETER
Niveau des Bacs
Equivalent Methane
Poids au Crochet
Densité
GEOSERVICES

USING RESISTIVITY MEASUREMENTS FOR THE DETECTION OF SALT BEDS

This example is based on the multitrack recording of the 1 930 to 1 940-metre interval of a well drilled in 1975 in a shaly series becoming evaporitic with salt beds.

1. TECHNICAL DATA

1.1. Drilling

- 13 3/8" casing shoe at 1 109 metres.
- Drilling with a 12 1/4" bit; mud density 1.16 (sea water-base); 2 400 l/min flow-rate for an average annulus volume of 65 l/m giving a theoretical lag time of about 50 minutes.

1.2. Logging

Geological control is ensured with the following equipment:

- Rate-of-penetration recorder (Speedograph + Rotomatic + Telescript) with remote depth marks inscription on the multitrack log
- Degasser and conductivity total gas detector
- Chromatograph for gas analysis
- H_2S detector
- Mud input and output densimeter
- Recorders for:
  - mud-pit level
  - pump rate
  - weight-on-hook
  - rotary speed
  - input and output temperature
  - mud resistivity

1.3. Recorded data and scales

Multitrack recording on a 280-millimetre-wide chart unwinding vertically from top to bottom as a function of time. The vertical scale is 7 min 30 s for 1 centimetre (between two horizontal lines).

The following logs are recorded on the multitrack chart:

- The depth marks are recorded every 50 centimetres in the right-hand margin of the chart.

- The total gas curve was recorded, but for an unknown reason the curve is reversed although the equipment operated quite normally when checked. The total gas curve is thus unusable.

- Mud-pit volume is recorded with a 10 m^3 full-scale measuring range; a large division corresponding to 1 m^3, a small one to 100 litres.

- Mud input and output density are recorded on scale 1, with a 0.75 to 1.75 range. A large division corresponds to 0.1, a small one to 0.01.

- Mud resistivity (or rather conductivity) is recorded on scale S3 with a conductivity range of 0 to 100 millisiemens/cm (resistivity from infinity to 0.1 ohm.metre), or S2 with a conductivity range of 0 to 50 millisiemens/cm (resistivity from infinity to 0.2 ohm.metre).

Another multitrack recorder was used for recording the input and output mud temperature, the weight-on-hook, the rotary speed and the pump flow rate. These data are not shown because the temperature measurements are not scaled and are thus unusable.

2. OPERATION DISCUSSION

A-B 04:50 to 06:00 hrs

Steady drilling from 1 930 to 1 936 metres with a rate of penetration of 15 min/m. A pipe connection was made at 1 934 m.

At pipe connection:

- the total gas curve, which should come back to zero, deflects to the right (return to zero with the curve reversed),

- the circulation tank volume increases about 3 m^3 (mud return from flow-line and mud circuit).

The unsteady decrease in mud volume observed while drilling is due to a leakage of the riser joint.

B-C 06:00 to 06:15 hrs

The mud conductivity shows an increase, going from 42 to 49 mS/cm @ 25°C [The measuring probe includes a built-in automatic temperature compensation circuit which brings the measured conductivity values to a 25°C reference for a sample temperature between 1 and 90°C]. Conversely, the mud resistivity (Rm) decreased from 0.238 to 0.204 ohm.metre.

To obtain the corresponding salinity of the mud filtrate, the mud filtrate resistivity (Rmf) must first be calculated using the Schlumberger GEN 7 chart:

Rm: 0.238	Rmf: 0.180	Salinity: about 36 g/l
Rm: 0.204	Rmf: 0.155	Salinity: about 45 g/l

These salinity values, estimated from the mud conductivity recording, are equivalent to those obtained using the classical mud filtrate measurement (41 to 50 g/l).

C 06:15 hrs

Drilling is stopped at geologist's request to circulate and confirm entry into salt.

C-D 06:15 to 06:50 hrs

Circulating

D 06:50 hrs

Drilling is resumed, salt having been found in the cuttings.

3. COMMENTS ON RECORDER FUNCTIONING

- A malfunctioning of the total gas detector (defective filament) gives an unusable reversed curve.
- The temperature curves, although correctly recorded, are not usable because the recorder scales were not properly noted on the chart.
- The other curves were properly recorded.

4. CONCLUSION

Using the resistivity measurement to detect entry into a salt bed was conclusive. The resistivity (or rather its reciprocal, the conductivity) started varying 50 minutes after entering the bed, which is in agreement with the theoretical lag time given.

CASE 7 CIRCULATING OUT A GAS KICK

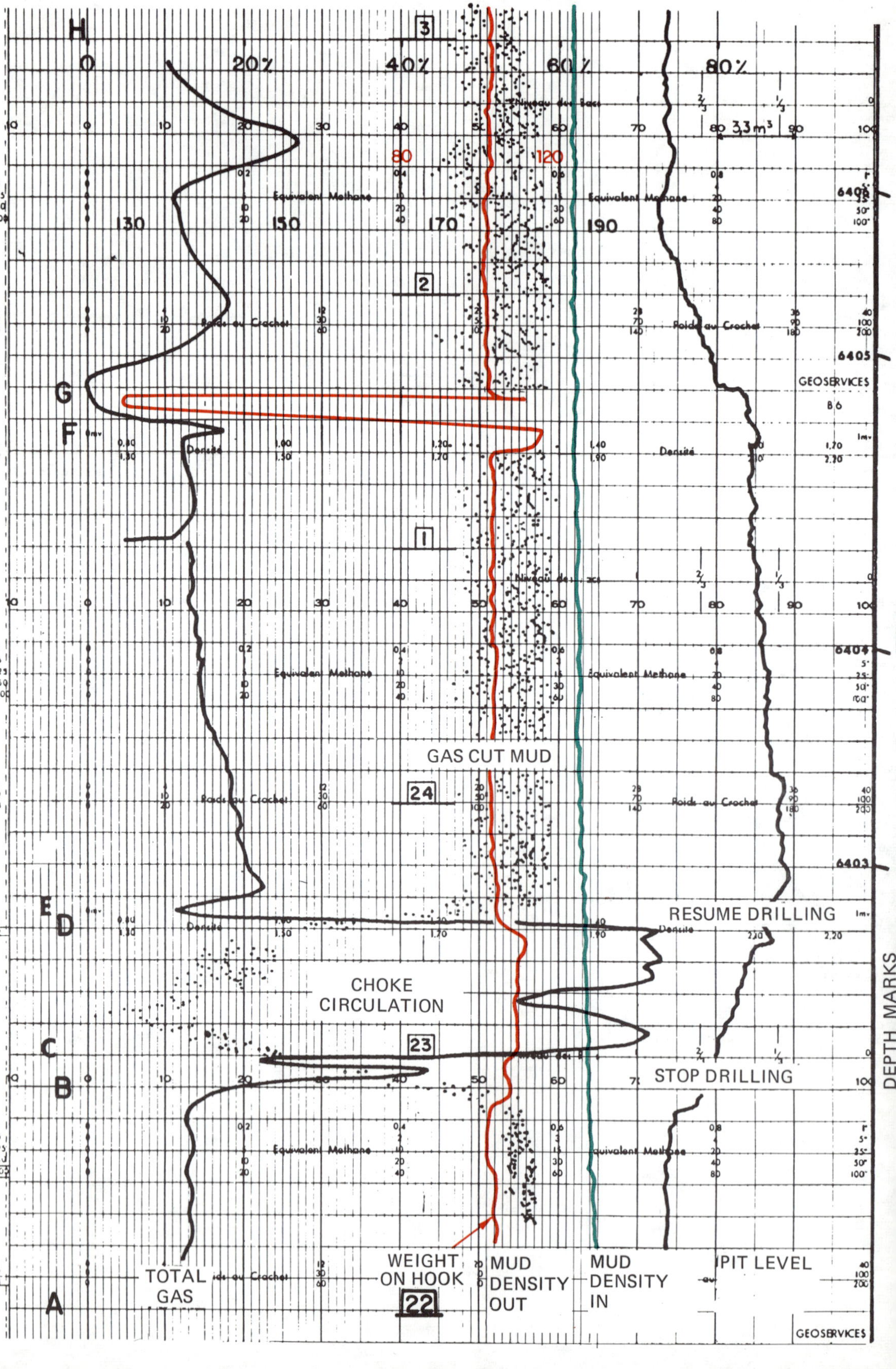

CIRCULATING OUT A GAS KICK

This example is based on the multitrack recording of the 6 402 to 6 406-metre interval of a deep well drilled in 1970 in a calcareous formation containing gas.

1. TECHNICAL DATA

1.1. Drilling

- 5 3/4" hole with a 500 l/min flow-rate at 6 400 metres
- The theoretical lag time is 2 hours and 30 minutes

1.2. Logging

Geological logging is ensured with the following equipment:

- Rate-of-penetration recorder with remote depth-mark inscription on the multitrack and chromatograph charts
- Degasser and conductivity total gas detector
- Chromatograph for gas analysis
- Mud-pit level recorder (Restor 3)
- Mud input and output densimeters
- Weight-on-hook recorder

1.3. Recorded data and scales

Multitrack recording on a 280-millimetre-wide chart unwinding vertically from top to bottom as a function of time. The vertical scale is 7 min. 30 s for 1 centimetre (between two horizontal lines).

- The depth marks are recorded every metre in the right-hand margin of the chart.

- The total gas curve is recorded on the S100 scale, corresponding to 100% full-scale methane equivalent in the analyzed gas-air mixture; a large division corresponds to 10%, a small one to 1%.

- Mud-pit volume variation is recorded with a 33 m^3 full-scale measuring range; a large division corresponds to 3.3 m^3, a small one to 0.33 m^3.

- Mud input and output density are recorded on a 1.25 to 2.25 scale. A large division corresponds to 0.1, a small one to 0.01.

- The weight-on-hook is recorded on a 0 to 200 ton scale. A large division corresponds to 20 t, a small one to 2 t.

2. OPERATION DISCUSSION

A-B 22:00 to 22:48 hrs

Drilling at a slow rate; the total gas curve shows a steady 13% background level; the mud-pit level is steady. The input mud density is stable at 1.9 while the output density values are somewhat scattered between 1.79 and 1.82.

B-C 22:48 to 23:00 hrs

A mud gain of 1.32 m^3 (4 divisions) is recorded. The total gas percentage starts increasing, and the output mud density shows a dramatic fall.

Drilling is stopped (quite noticeable on the weight-on-hook curve: + 6 t), and the blowout preventers are closed before the arrival of the gas slug at the surface.

C-D 23:00 to 23:30 hrs

Choke circulation to evacuate the gas slug. During this operation we see that

- the mud-pit level steadily increases until 23:28 hrs, and then shows a slight decrease,

- the total gas exceeds 70% methane equivalent in the analysed gas-air mixture,
- the output mud density, with quite scattered values, rapidly reaches a minimum of 1.30 and then increases steadily.

D-E 23:30 to 23:35 hrs

Choke circulation is stopped and the BOPs opened.

E-F 23:35 to 01:25 hrs

Drilling resumed:

- The total gas percentage steadily decreases from 22 to 13%.
- The mud-pit level goes steadily down.
- The output mud density values are quite scattered (1.75 to 1.85); the mud still contains a high proportion of gas.

F-G 01:25 to 01:35 hrs

Pipe connection, clearly shown on the total gas and weight-on-hook curves.

G-H 01:35 to 03:00 hrs

Drilling in progress. Notice that the mud still contains gas long after the arrival of the gas slug. The total gas curve shows a 27% peak at 02:30; it is most probably due to the recycling of the gas slug which led to choke circulation: the 3-hour lapse checks with the lag time (2 hrs 30 min) added to the surface-to-bottom mud-travel time inside the pipes (about 30 minutes).

3. CONCLUSION

This is a good example of the sort of help the multitrack recordings may offer when the mud is highly gas-cut, because it allows:

- anticipation of a gas kick while drilling,
- monitoring of the input and output mud density while maneuvering.

In this typical case, the time between the first indication of the arrival of gas (scattering of the output mud density) and its actual arrival at the surface was about one hour. The start of the gas slug decomposition process, responsible for the pit-volume increase, left a margin of about 10 minutes.

CASE 7

CASE 8

GAS KICK RESULTING IN FORMATION FRACTURATION AFTER CLOSING THE BOP

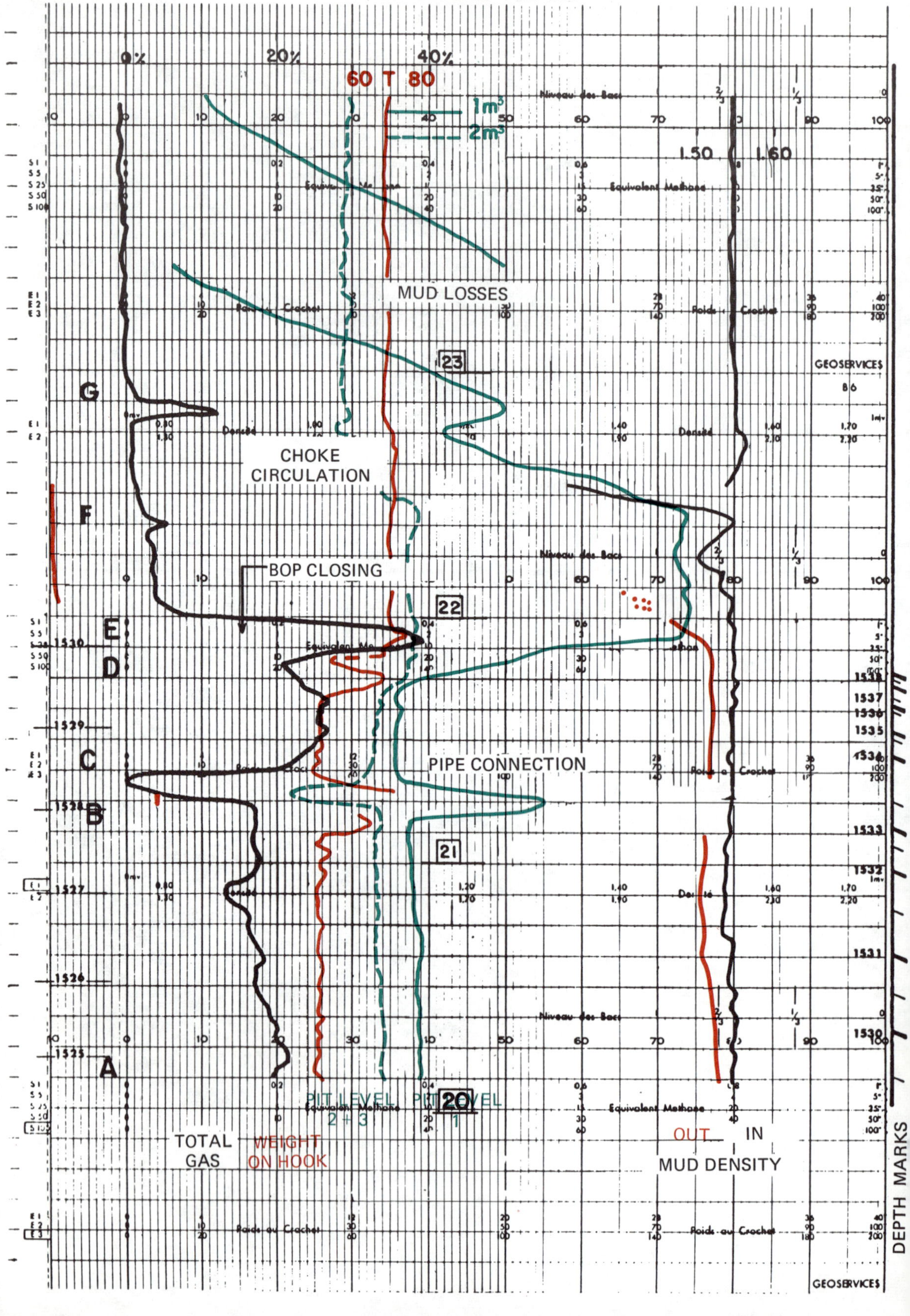

GAS KICK RESULTING IN FORMATION FRACTURATION AFTER CLOSING THE BOP

This example is based on the multitrack recording of the 1 529- to 1 538-metre interval of a well drilled in 1971 in a deltaic formation.

1. TECHNICAL DATA

1.1. Drilling

- 20" casing shoe at 404 metres
- Hole diameter: 18 1/2" down to 1 386 metres
- Drilling with 12 1/4" bit (reduced hole)
- Mud flow-rate: 2 600 l/min
- T.D. 1 538 metres
- Theoretical lag time 1 hr 30 min

1.2. Logging

Geological logging is ensured with the following equipment:

- Rate-of-penetration recorder (Speedograph + Rotomatic) with remote depth-mark inscriptions on the multitrack and chromatograph log
- Degasser and total gas detector
- Chromatograph for gas analysis
- Mud-pit level recorder (Restor 3)
- Mud input and output densimeter
- Weight-on-hook recorder

1.3. Recorded data and scales

This multitrack log was recorded on a 280-millimetre-wide chart unwinding vertically from top to bottom as a function of time. The vertical scale is 7 min. 30 s for 1 centimetre (between two horizontal lines).

The following logs are recorded on the multitrack chart:

- The depth marks are recorded every 50 centimetres in the right-hand margin of the chart.

- The total gas curve is recorded on the S100 scale, corresponding to 100% full-scale methane equivalent in the analyzed gas-air mixture; a large division corresponds to 10%, a small one to 1%.

- The mud-pit volume variation is shown on two separate curves:
  . Volume variation of pit #1 (input), with a 10 m^3 full-scale measuring range; a large division corresponding to 1 m^3, a small one to 100 litres.
  . Total volume variation of pit #2 plus pit #3, with a 20 m^3 full-scale measuring range; a large division corresponding to 2 m^3, a small one to 200 litres.

- The mud input and output density is recorded with a 0.75 to 1.75 range. A large division corresponds to 0.1, a small one to 0.01.

- The weight-on-hook is recorded on a 0 to 200-ton scale. A large division corresponds to 20 t, a small one to 2 t.

OPERATION DISCUSSION

A-B 20:07 to 21:10 hrs

Drilling with an average rate of penetration of 20 min/m. At 21:00 hrs, the rate increases to 10 min/m.

The only noticeable facts are
. a slight decrease in mud-pit level due to filling of the drilled section,
. a rather high gas background level: 13 to 22%.

B-C 21:00 to 21:20 hrs

Pipe connection giving rise to

. an increase of the weight-on-hook when pulling off-bottom, followed by a sharp decrease when putting the drill-string on slips,
. an increase in mud-pit #1 level due to the emptying of the flow-line, and a decrease in mud-pit #2 and #3,
. the return to zero of the total gas curve, with a delay corresponding to the transit time from degasser to detector (3-4 minutes).

C-D 21:20 to 21:45 hrs

Drilling with a faster drilling rate (5 to 7 min/m) (reservoir top at 1 532 metres).

D-E 21:45 to 21:55

3.8 m^3 mud gain in pit #1 and 600 l in pits #2 and #3 in 10 min.

Total gas reaches 38%.

At 21:55 hrs, the blowout preventers are closed before the arrival of the gas slug at the surface. The total gas curve decreases sharply since the degasser is not fed anymore.

F-G 22:20 to 22:50

Choke circulation to transfer the gas.

Starting at 22:50 hrs, significant level decrease in pit #1, due to mud losses following fracturation of the formation. The fracturation was due to overpressure caused by the gas rising in the annulus (see appendix).

3. CONCLUSION

In this example, the technical quality of the logs is good, the total-gas curve is steady and the return to zero is correct. However, one must remember that during choke circulation some parameters are not recorded, because the sensors placed in the mud-return line are not fed anymore. Here, the total gas and the output mud density curves could not be recorded. A device for connecting the sensors during choke circulation is available, but it was not installed here.

APPENDIX
PRESSURE-VARIATION MODES IN THE CASE OF A GAS SLUG MIGRATING UP A CLOSED ANNULUS

When a well is being closed, the annulus volume is constant. Thus, the gas slug volume is constant. Applying Mariotte's law, PV = constant, the pressure of the slug stays the same whatever the slug position in the annulus.

We then deduce that

- the surface pressure is equal to the gas slug pressure (pore pressure) less the hydrostatic pressure of the mud column above the gas slug,
- the bottom pressure is equal to the gas slug pressure plus the hydrostatic pressure of the mud column between the gas slug and the hole bottom,
- the pressure applied to the formation at any depth as the gas rises towards the surface is equal to
  - . the mud column hydrostatic pressure plus the surface pressure for a point located above the gas slug,
  - . the gas slug pressure plus the pressure of the mud column between the slug and the point of interest for a point located below the gas slug.

The evolution of the pressures is illustrated in the accompanying schematic representation. It shows that, when the slug is at the surface, the head pressure is 360 bars*, the pressure applied to the formation at 1 500 metres is already 540 bars (equivalent mud density 3.6) and that it reaches 720 bars at the bottom of the hole.

In normal operations, the gas slug should not be allowed to rise to the surface with the well closed, because the resulting overpressure may fracture the formation (or the casing), as in the previously discussed example. The dangers are numerous: internal eruption, pressure rise in the casing annulus, migration of gas behind the casing and surface blowout (cratering), leaks or blowout when the gas pressure reaches or exceeds the nominal pressure rating of the surface equipment.

* To simplify, and taking into account the accuracy of the field instruments which would be used on a drilling rig, we consider 1 bar equal to 1 kg/cm^2.

RISE TO SURFACE OF 100 LITRES OF GAS IN A WELL CLOSED AT THE TOP

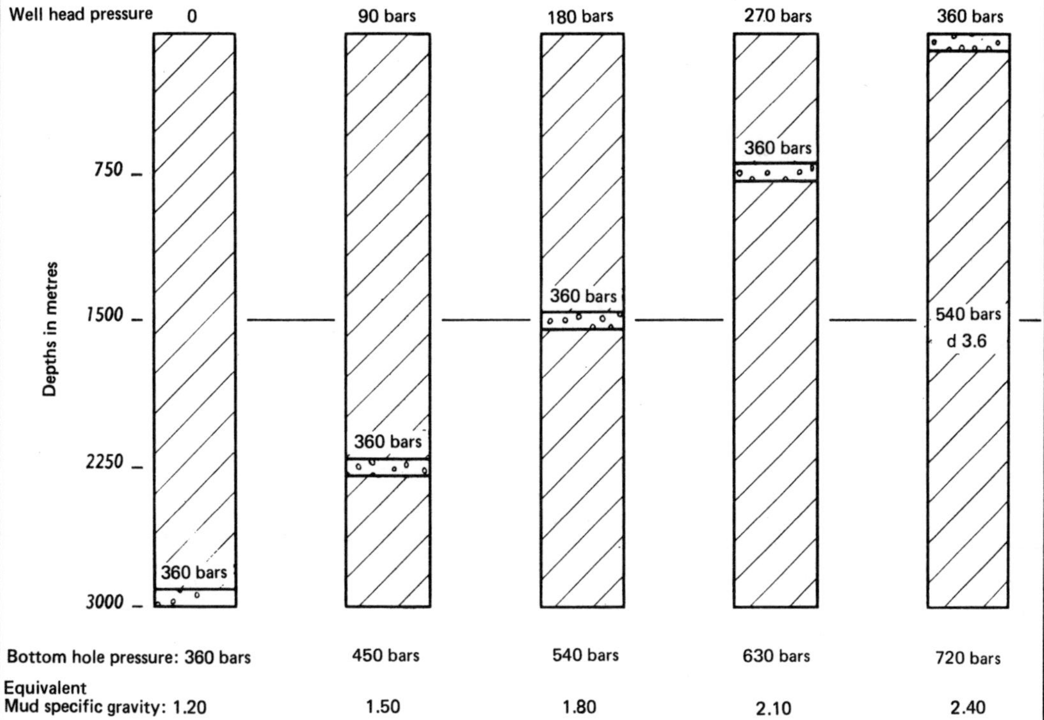

At 3 000 m with a 1.2 specific gravity mud, the gas pressure is 360 bars. As the gas rises to the surface, without expansion because the well is closed, it remains under a 360-bar pressure. On arrival at the surface, this pressure, combined with the hydrostatic pressure of the mud, creates a downhole pressure of 720 bars; this corresponds to a mud specific gravity of 2.40. At 1 500 m the pressure is 540 bars; this corresponds to a mud specific gravity of 3.60.

The rise of the gas by diffusion in the mud column may be estimated roughly at 300 m per hour.

WATER FLOWING WELL

This example is based on a log recorded in 1973. The well produced water from a dolomitic bed which was expected to produce gas.

1. TECHNICAL DATA

1.1. Drilling

- 9 5/8" casing shoe at 2 921 metres
- 8 1/2" hole
- Salt-saturated mud, density 1.25, viscosity 35-37, flow rate 1 965 l/min;
- Circulated mud volume: 193 m^3, i.e.:
  . in the well: 111 m^3
  . in the pits: 80 m^3
  . in the flow-line: 2 m^3
- Theoretical lag time: about 40 minutes.

1.2. Logging

Geological control is ensured with the following equipment:

- Rate-of-penetration recorder with remote depth marker
- Pump stroke counter
- Chromatograph for gas analysis
- Total gas detector
- Mud output densimeter
- Mud-pit level recorder

The data from the three last instruments together with the depth marks were logged on a multitrack recorder

1.3. Recorded data and scales

The 264-millimetre-wide chart unwinds vertically from top to bottom as a function of time. The spacing between two horizontal lines corresponds to 2 min 30 sec.

The chart is divided into three tracks:

Track 1: Total gas curve

The track is 128 millimetres wide, has 10 divisions, each with 5 sub-divisions. Four different scales are available:

- 0-25 (1:1 signal attenuation) 10 units full-scale deflection
- 0-100 (4:1 signal attenuation) 40 units full-scale deflection
- 0-250 (10:1 signal attenuation) 100 units full-scale deflection
- 0-1000 (40:1 signal attenuation) 400 units full-scale deflection

For the last scale, each gas unit is equivalent to 50 ppm of gas in the analysed mixture.

Track 2: Output mud density

The track is 64 millimetres wide, has 4 divisions, each with 5 sub-divisions. In this example, each sub-division represents a density variation of 0.01, and the track covers the density range 1.16 to 1.36.

Track 3: Mud-pit level

The track is 64 millimetres wide, has 4 divisions, each with 5 sub-divisions. Each sub-division equals 18 barrels, i.e. 2.86 m^3.

2. OPERATION DISCUSSION

A 06:33 hrs

The pit volume increases 1 small division, equivalent to a mud gain of about 3 m^3 in 3 minutes.

Note: the densimeter is out of order until 07:20; when repaired it indicates d = 1.25.

From 06:33 to 07:30 hrs: drilled 4 feet. During that period the pit level seems to increase slightly.

B 07:00 hrs

Small gas show. The maximum value (1 300 ppm or 0.1%) is attained at 07:27 hrs. This gas index seems to correlate with the mud gain observed at 06:35 hrs.

C 07:30 hrs

Drilling and then circulation are stopped to observe the well. In 20 minutes, the pit level increases one and a half sub-division, i.e. about 4 m^3. As the flow-line volume is only 2 m^3, its emptying does not fully explain this gain. The formation is thus flowing about 2 m^3.

When the circulation is resumed (7:42 hrs), the level decreases less than one small division, which agrees with the 2 m^3 required to fill the flow-line.

D 07:50 hrs

Beginning to weight the mud, which results in a steady increase in mud-pit level (baryte addition + inflow).

E 08:20 hrs

38 minutes after circulation was resumed, arrival of a small gas slug (10 000 ppm, i.e. 1%) and decrease in density from 1.25 to 1.20, without any scattering of the density values, which seems to indicate the arrival at the surface of a water slug.

From 08:20 till 09:38 hrs: Still weighting the mud; the density decreases slightly and reaches 1.26.

F 09:38 hrs

Stopped circulating; 3.5 m^3 mud gain in 15 minutes. As before, the gain is larger than the emptying of the flow-line, which indicates that the formation produces about 1.5 m^3.

G 09:53 hrs

Resumed circulation. In spite of the level decrease due to the flow-line filling, the mud-pit level stays half a small division (1.5 m^3) above what it was before circulation was stopped.

H 10:33 hrs

The density decreases again from 1.27 to 1.22, and there is a slight gas show 37 minutes after circulation is resumed.

I 10:33 to 12:45 hrs

Still weighting the mud. A density of 1.31 is reached, and the pit level stabilizes. The total gain since the first inflow is about 33 m^3 (baryte + inflow).

From 12:45 till 13:33 hrs: circulating, no baryte addition.

J 13:33

Stopped circulating. After circulation is stopped, the gain is still larger than the flow-line volume, and 40 minutes later there is again a decrease in density together with a small gas show.

The phenomenon will be observed each time after the circulation is stopped.

3. COMMENTS

3.1. Log quality

The overall log quality is good and all required information is properly entered.

The total gas curve scale has been changed several times (at about 3/4 of full-scale deflection).

The following should also be noted:

- 2 failures (electrical?) of the detector, at 11:48 and 12:00 hrs. These failures appear to be coming from a bad contact in the detection circuit.
- No adjustment or zero check was made while this log was recorded.
- Densimeter failure; repaired at 07:20 hrs.

The lack of sensitivity of the pit level measurement is regrettable as only the large gains or losses (larger than 1 m^3) may be detected.

3.2. Log interpretation

The insensitivity of the mud-pit level measurement makes it difficult to observe certain phenomena and renders the interpretation of the recorded information very difficult.

In spite of this, there cannot be any doubt about the nature of the phenomenon. The following points lead to the conclusion that the inflow was salt water containing some dissolved gas:

- Low gas indexes; even the recycled-gas shows are small when circulation is resumed (remember that an inflow of 100 litres of gas at 3 000 metres and a pressure of 360 bars would give 36 m^3 of gas at the surface). In each case, the gas show was proportional to the time the circulation was stopped and thus to the amount of water produced by the formation.

- Each time the fluid from the producing zone arrives at the surface after the circulation is stopped, a decrease in mud density is observed, corresponding to the arrival of the water at the surface (it is unfortunate that no sampling for salinity measurement was made at that time).

- The densimeter readings do not show the scattering which is typical of gas-cut mud.

- The production tests done at this level indicated an inflow of 40 m^3/hr salt water @ 305 g/l with a density of about 1.2.

3.3. Calculation of the amount of added baryte

X = water volume, density d_e = 1.2
Y = baryte volume, density d_b = 4.0
V1 = initial mud volume (d1 = 1.25): 193 m^3
V2 = final mud volume (d2 = 1.31)

The mud volume increase was 11.5 divisions = 11.5 x 2.86 = 33 m^3

$V2 = 193 + 33 = 226 \text{ m}^3$

$X + Y = V2 - V1 = 33 \text{ m}^3$ (1)

$(X\, d_e) + (Y\, d_b) = (V2\, d2) - (V1\, d1)$ (2)

From (1), X = 33 - Y, thus:

$(33 - Y)\, 1.2 + 4\, Y = (226 \times 1.31) - (193 \times 1.25)$

then

$2.8\, Y = 15.2 \qquad Y = 5.43 \text{ m}^3$

and

$X = 33 - 5.43 = 27.57 \text{ m}^3$

The water inflow was about 27.5 m^3, and the volume of added baryte was 5.5 m^3, i.e. 22 tons.

Notice that the cumulated gain of 27.5 m^3 in 5 hours, i.e. 5.5 m^3/hr in average, agrees with the gain of about 2 m^3 in 20 minutes observed at the beginning of the kick, and with that of 1.5 m^3 in 15 minutes recorded when the circulation was stopped between 09:00 and 10:00 hrs.

USING TEMPERATURE MEASUREMENTS AND "d" EXPONENT CALCULATIONS FOR THE DETECTION OF UNDERCOMPACTED SHALES

In deltaic regions, there is a constant need for better means of detecting the undercompacted shaly formations and associated high-pressure zones. This example illustrates the use of the temperature measurements and "d" exponent calculation to detect undercompacted formations. The cases discussed are from wells drilled in 1973-1974.

1. DATA USED FOR THE DETECTION

1.1. Drilling control and geological logging

- "d" and/or "dcs" exponent calculation ("dcs" exponent is corrected for mud and bit wear effects). This calculation requires the following parameters to be recorded:
  - drilling speed or rate of penetration
  - weight-on-hook
  - bit rpm
- Input and flow-line mud temperatures
- Output mud density
- Shale density taken from the cuttings
- Gas indexes

1.2. Electrical logging

- Sonic (shale Δt)
- Density
- Resistivity

2. SUPPORTING DOCUMENTS

To better visualize the contribution of each detection method, a synthetic log was drawn for each well. The following information was plotted versus depth:

- A succinct stratigraphic cross-section showing the reservoir zones and the fluid they contain

- The values taken from the electrical logs (shale Δt, density and resistivity)

- The calculations* and measurements from the drilling control and geological logs:

  . mean "flow-line" temperature gradient (°C/100 metres)
  . shale density (cuttings)
  . output mud density
  . uncorrected "d" exponent
  . "dc" exponent corrected for mud density
  . "dcs" exponent corrected for mud density and bit wear

3. RESULTS INTERPRETATION AND COMMENTS

3.1. Well A (Figure 2)

The "d"exponent could not be calculated for this well because of a malfunctioning rate-of-penetration recorder. An attempt to calculate a pseudo-"d" exponent was made using a very approximate rate of penetration.

According to the electrical logs (shale Δt and density), the top of the undercompacted zone is around 2 150 metres.

Using the geological logs, the following changes were observed:

- The mud mean temperature gradient at the flow-line changes from 2°C/100 m to 3.5°C/100 m around 2 065 metres, i.e. 85 metres before entry into the undercompacted zone.

- The shale cuttings density curve does not show any significant change (possible undercompaction top around 2 120-2 140 metres ?).

* The calculation of the "d" exponent is given in figure 1. "dc" is obtained using RT (rate of penetration at time T), and "dcs" using Ro (rate of penetration corrected for bit wear).

- The "pseudo-d exponent" curve is difficult to use due to the numerous bit changes (change of direction of the curve around 2 100-2 110 metres) and the crossing of numerous sand layers.

In this well, only the temperature gradient of the mud has increased in a significant way, 85 metres before the entry into the undercompacted zone.

3.2. Well B (Figure 3)

According to the electrical logs (shale density and resistivity), the top of the undercompacted zone is around 2 100 metres (or 2 140 metres if we take the shale Δt).

The mean temperature gradient of the mud in the flow-line increases from 2 to 4°C/100 m at around 1 970 metres, i.e. at about 130 metres above the top of the undercompacted zone. A second increase from 4.5 to 7°C/100 m is monitored at around 2 000 metres.

The "d" and "dc" exponent curves agree and show the entry into the undercompacted zone around 2 090-2 100 metres.

As in well A, the first parameter to react is the temperature of the mud in the flow-line, which increases 130 metres before the top of the undercompacted zone, while the "d" exponent shows the top of that zone to be at the same depth as that obtained from the electrical logs.

3.3. Well C (Figure 4)

According to the electrical logs (shale density and Δt), the top of the undercompacted shales is at around 1 740 metres.

The mean temperature gradient of the mud in the flow-line shows two increases at 1 640 and 1 645 metres i.e. respectively 100 and 75 metres above the top of the undercompacted zone.

The density of the shale from the cuttings starts decreasing at 1 760 metres.

Of the three "d" exponent curves, only the "dcs" curve, corrected for mud density and bit wear, shows the top of the undercompacted zone at around 1 750 metres; both the other curves react later. This illustrates the advantage of calculating a corrected "d" exponent.

3.4. Well D (Figure 5)

According to the electrical logs (shale density and Δt), the top of the undercompacted shales is between 1 550 and 1 560 metres.

The temperature gradient of the mud in the flow-line increases from 0.8 to 1.5°C/100 m at around 1 510 metres i.e. 40 metres above the top of the undercompacted zone.

The shale density from the cuttings decreases from 2.32 to 2.15 at around 1 575 metres.

Of the three "d" exponent curves, only the "dcs" curve corrected for mud density and bit wear reacts at around 1 560 metres; both the other curves react later, at around 1 590 metres, i.e. 40 metres after entry into the undercompacted zone.

3.5. Well E (Figure 6)

The electrical logs show that there are two compaction changes in this well, one around 1 650-1 670 metres where the shale Δt and density no longer change with depth, and another at around 1 800 metres where we have the top of a better defined undercompacted zone.

The interpretation of the mud logs gives less conclusive results:

- The temperature gradient of the mud in the flow-line increases from 1.8 to 2.2°C/100 m at around 1 690 metres and from 2.2 to 3°C/100 m at 1 850 metres, i.e. respectively 150 and 20 metres above the top of the better defined undercompacted zone.

- The shale density from the cuttings reaches a maximum of 2.35 at around 1 665 metres, and then steadily decreases to reach 2.25 at 1 860 metres. It remains stationary from 1 860 to 1 930 metres, where it starts decreasing again.

Of the three "d" exponent curves, only the "dcs" curve corrected for mud density and bit wear correlates, although not very closely, with the electrical log data: the "dcs" exponent increases down to 1 660 metres, remains constant from 1 660 to 1 810 metres, and then decreases below 1 870 metres (between 1 810 and 1 863 metres: 2 sand beds are encountered, causing the "dcs" exponent to decrease as expected). The top of the undercompacted zone cannot be located with precision.

4. CONCLUSION

The examples discussed illustrate the utility of the geological logs in the detection of undercompacted shales.

When considered as "warning indicators", the data from the mud logs may be classified in order of decreasing interest as follows:

- The mud temperature measurement at the flow-line seems to be most interesting, because the mean temperature gradient shows a significant increase 40 to 150 metres before entering an undercompacted zone. It is a good warning indicator, as it allows the detection of an undercompacted formation before it is actually entered. The only delay in obtaining the information is due to the lag time.

The examples studied come from offshore wells in warm seas; in cold seas and at great water depth, the Calorimud is not usable (due to the heat exchange in the riser).

- In these examples, the "dcs" exponent corrected for mud density and bit wear gives the most representative curve. However, experience shows that very often the "dc" exponent corrected for mud density only is as useful, because the bit wear calculation is too empirical for the correction to be really meaningful. The "dc" or "dcs" exponent shows entry into undercompacted shales while actually drilling through the transition zone. Thus, the information is received later than with the temperature measurements. Another point to remember is that a decreasing "dcs" exponent does not necessarily mean an entry into undercompacted shales, but could also indicate a change in lithology (sand beds in particular).

  A more representative and precise "d" exponent curve may be obtained by choosing carefully the data points (levels at which one is certain to be drilling through shale only).

- Measurement of the shale density from the cuttings may offer interesting indications, but the information is not immediately available. It is delayed not only because of the lag time, but also due to the time necessary to perform the analysis. Furthermore, the consistency of the shales, very often disaggregated by the drilling mud, may make the measurement doubtful or even impossible.

DRILLING THROUGH AN EVAPORITIC SERIES

This example is based on the 3 215- to 3 335-metre interval of a well drilled in 1976 through an anhydritic and saliferous series.

1. TECHNICAL DATA

- Depths are in metres
- 9 5/8" casing shoe at 2 425 metres
- Drilling in progress at 3 215 metres; 8 1/2" bit
- Salt saturated water-base mud; density 1.2

2. AVAILABLE LOGS & DOCUMENTS

Figure 1 is the 1/500 scale log recorded at the well site. It shows the following information, from left to right:

- Calcimetry (or rather carbonate content): percentage of carbonates contained in the cuttings.
- Cutting contents in percentages.
- Lithological cross-section drawn by the field geologist as drilling progresses.
- Depth scale.
- Rate of penetration in minutes per metre, scale 2: 0 tc 100 min/m full scale, recorded every half-metre. The drilling parametres are also indicated on this track.
- Mud data: mud characteristics (density, viscosity, filtrate) and output temperature. On this track are indicated the transit time of the mud from the bottom to the surface (lag time) and the hole deviation measured while drilling (Totco).
- Hydrocarbon shows: composition in percent of the mixture coming from the degasser installed in the flow-line close to the riser. Measurements were also made of the actual quantity of gas contained in the mud: these punctual analyses were made after complete degassing of the mud under vacuum; the results are recorded on the track.
- Description of the cuttings collected at the shale-shakers by the field geologist.

Figure 2 is a composite log made with part of the field log and the electrical logs: Gamma-Ray, Caliper, Formation Density, Neutron and Sonic.

2. OPERATION DISCUSSION (Figure 1)

3 215 to 3 296 metres

Drilling is progressing with a highly variable rate of penetration. At 3 220 metres, the rate of penetration which was varying between 8 and 16 min/m starts slowing down, reaching values in excess of 20 min/m starting at 3 221.5 metres, and even 40 min/m at 3 223 metres. At 3 231 metres, a sharp increase in speed is clearly shown by the curve which stays below 10 min/m down to 3 246.5 metres, except for a section around 20 min/m between 3 241 and 3 242 metres. Drilling slows down again with a rate of penetration varying between 20 and 30 min/m down to 3 258 metres. From that point down, the ROP is more scattered and shows frequent variations between 9 and 25 min/m. Drilling seems to slow down in the last 4 metres.

In this interval, the cuttings are mostly made of transparent salt. The anhydrite percentage is non-negligible but generally small, seldom reaching more than 30% of the cuttings; the variation in cutting percentages shows a certain correlation with the variations in the penetration rate. At the top of the interval, light marls represent 60 to 70% of the cuttings; they rapidly diminish, to remain only as traces, except between 3 245 and 3 252 metres. The 30% carbonate content at the top of the example rapidly decreases to 10% and then to lesser values.

Hydrocarbon shows: traces of gas confirmed by the reference degasser.

Drilling is stopped at 3 296 metres to change the bit.

3 296 to 3 335 metres

When drilling is resumed with a new bit, the slowdown noted before drilling stopped is not confirmed. The rate of penetration is highly variable down to 3 300 metres. A 3.5-metre section is then drilled at 20 to 30 min/m, followed by a zone with a faster penetration rate: 10 to 20 and then 5 to 10 min/m, down to 3 320 metres. The drilling speed then decreases regularly and stabilizes at around 25 min/m between 3 326.5 and 3 335 metres which is the bottom of the section considered here.

The cuttings are made of the same components as in the previous section: salt is strongly dominant down to 3 320 metres. At that depth the anhydrite reaches 60%, to decrease slightly afterwards. The complement is then mostly made of a grey, saliferous, anhydritic marl. In this zone, a certain correlation is observed between the percentage of salt and anhydrite and the rate of penetration, although this does not seem to be true in the last ten metres of the interval.

The carbonate content is very low; still no hydrocarbon show.

3. COMMENTS

The lithological cross-section drawn by the field geologist shows:

- massive anhydrite beds
- thinner anhydrite layers
- beds of salt mixed with marls and dispersed anhydrite

It is obvious that the geologist mainly referred to the rate-of-penetration curve to draw the lithological cross-section: the fast and slow penetration rates respectively correspond to saliferous beds and anhydrite beds.

A more detailed analysis using the electrical logs (Fig. 2) confirms this interpretation while making it more precise. This more detailed analysis allows the association of certain slowdowns to shaly (marly) beds, which is not apparent in the data collected while drilling alone.

4. CONCLUSION

In this example, the evaporitic series is almost perfectly described using the cutting analysis, thanks to the well adapted salt-saturated mud. Identical results would have been obtained using an oil-based mud, and it is likely that the hole would have been in better shape than is shown on the caliper log which indicates that the salt beds are caved.

The interpretation of an evaporitic series is mainly based on the rate-of-penetration analyses which indicate quite accurately the passages through alternate formations. In Figure 3 the formation density and the rate of penetration logs were superimposed after depth-correction. The inverted caliper curve is also shown. The three curves correlate remarkably well, and so would the sonic log. If one admits, as a first approximation, that the density curve deflection to the left (low density) and to the right (high density) respectively correspond to salt and anhydrite, then the faster (salt) and slower (anhydrite) penetration rates almost perfectly correlate with the density curve. Some levels are even more remarkable:

- the 3 230- to 3 255-metre interval
- the salt-anhydrite contrast between 3 294 and 3 297 metres
- the thinner salt layers at 3 308.5, 3 314 and 3 327 metres
- the anhydrite beds at 3 277 and 3 313 metres

In some sections, when the hole presents large caves, it appears that the vertical definition given by the rate of penetration measured every half-metre is better than that of the elecrical logs run in this type of formation.

When drilling with a saturated saltwater-base mud, using the penetration rate allows an interpretation of the formation that cannot be made by examining only the cuttings, because of the multiple parasitic effects which affect them: cavings, unavoidable partial dissolution of the salt, deterioration and dispersion of the anhydrite and saliferous shales. If an oil-base mud is used, speed changes might not be as obvious, but the cuttings will give a better picture of the formation because they are not submitted to dissolution phenomena. Penetration-rate contrasts will also be highly attenuated when drilling with a diamond bit.

Fig. 3

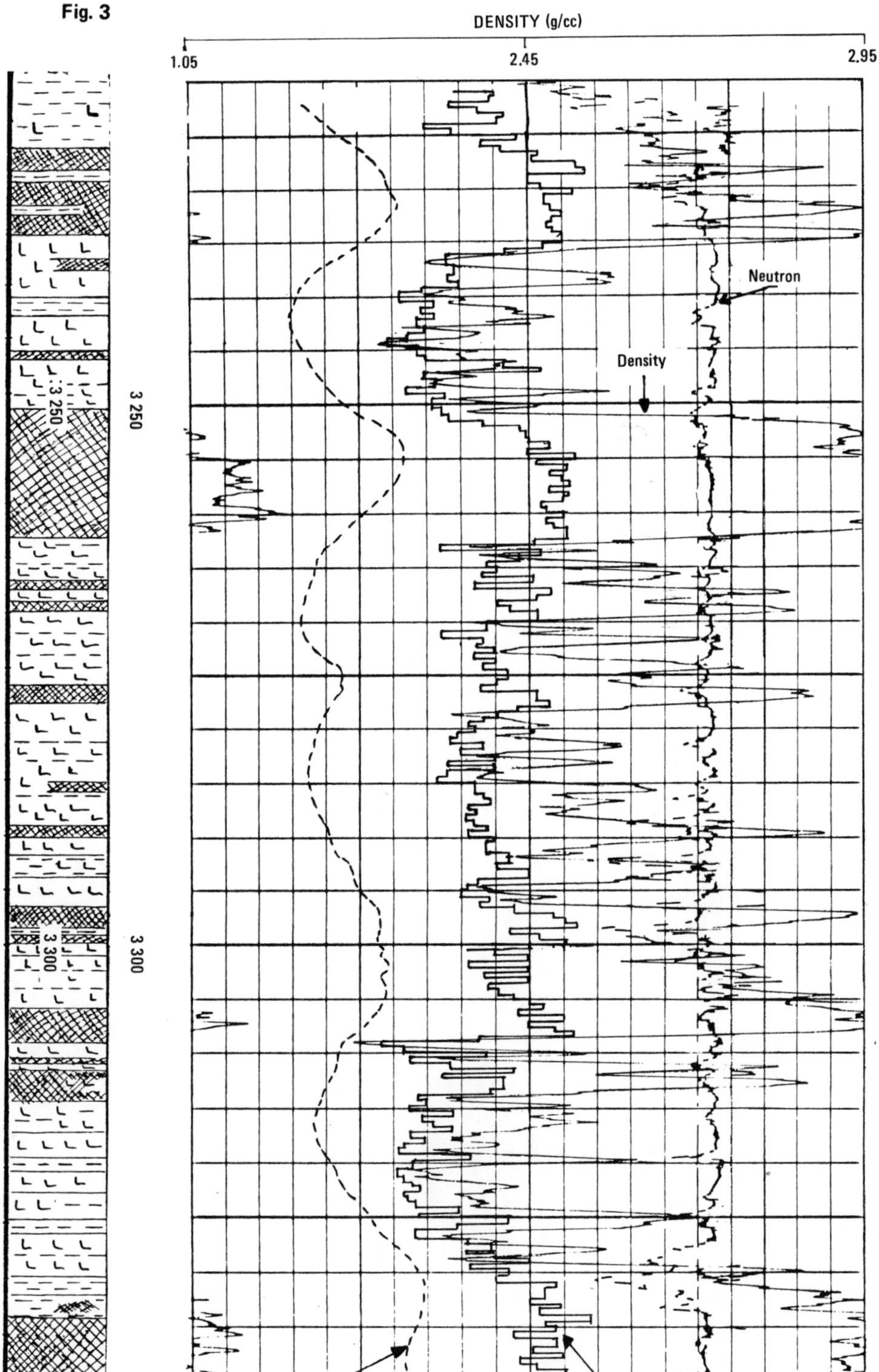

DENSITY (g/cc)
1.05
2.45
2.95
3 250
3 300
Neutron
Density
Reversed caliper log
Penetration rate
Caving
Drilling
Fast
Slow

DRILLING THROUGH A SHALY SAND SERIES WITH INTERBEDDED SALT LAYERS BECOMING MASSIVE SALT

This example shows the 8 500 to 8 850-foot interval of a well drilled in 1976 in a predominantly shaly sand series turning gradually to massive salt.

1. TECHNICAL DATA

- Depths are given in feet.
- Bit size is 8 1/2".
- Drilling mud is saltwater-base mud: in the upper section the salinity is 78 000 ppm with a density of 1.58. Below 8 774' oil-base mud is used.
- Drilling is just resuming with a new bit at 8 500'.

2. AVAILABLE DOCUMENTS

The well description log, drawn at a 1/500 scale, is shown in figure 1 and includes the following information, from the left to the right:

- Cutting percentage
- Depth
- Rate of penetration in minutes per foot, scale 1 = 0 to 50 min/ft full scale
- Mud data: salinity in parts per million (ppm)
- Gas detector curve giving, in percent of equivalent methane, the amount of gas contained in the analyzed gas-air mixture coming from the degasser installed in the return flow-line
- Cutting description done by the well-site geologist.

The interpretative lithological cross-section of the well was not made at the well-site.

Figure 2 is a composite log made up of some of the mud logs and the electrical logs: Gamma-Ray, Caliper, Neutron, Formation Density and Sonic.

3. OPERATION DISCUSSION (Fig. 1)

From 8 500 to 8 605 feet:

The rate of penetration, which is 5 min/ft at the beginning of this example, shows the following extreme variations:

- 3 min/ft from 8 525 to 8 535 ft
- 9 min/ft at 8 565 ft
- 4.5 min/ft at 8 580 ft
- 17 min/ft at the end of the interval

Drilling is then stopped to change the bit.

During this bit run, the cutting samples show, in slightly variable proportion:

- sandstone with no apparent porosity, representing less than 50% of the cuttings. At 8 590' appearance of coarse sand;
- the complement to 100% of the cuttings is mostly made of silty shale with quartz grains.

The gas curve which was first close to zero, shows an increase starting at 8 525', reaching 3.5% at 8 545'; the measured gas then decreases to its original level of 0.1% when drilling is stopped.

From 8 605 to 8 774 feet

After drilling is resumed with a new bit, the rate of penetration goes from 9 to 2 and then 1 min/ft. This fast drilling rate is maintained down to 8 770' with only a slight decrease in the rate between 8 725' and 8 730'.

The cuttings do not show any significant changes. The sandstone percentage seems to increase until 8 690 feet, where it reaches 65% of the cuttings; traces of anhydrite are also noticed. This percentage then sharply decreases and shales and siltstone become predominant (60 to 85% of the cuttings).

The gas curve remains almost at zero down to 8 680', then increases in steps, reaching 10% at 8 774'.

The salt content of the mud gradually increases from 78 000 ppm to 90 000 ppm.

At 8 770' salt starts showing in the cuttings.

Drilling is stopped at 8 774', the bit and drill pipes being stuck.

From 8 774 to 8 850 feet:

After freeing the drill string, the electrical logging operation takes place. The presence of salt having been confirmed, drilling is resumed after changing the water-base mud to oil-base mud.

From the beginning, the rate of penetration is lower than that of the previous bit run: from 2 to 5 min/ft. Cutting observation first indicates the persistence of sand and shale, quickly replaced by anhydrite and then by 80% massive salt beginning at 8 820'.

There are no more signs of gas; the gas detector curve is almost at zero.

4. COMMENTS

The interpretative lithological cross-section of the well was not drawn by the field geologist. No mention of salt or particular remarks having been made before 8 770', one may suppose that it was not detected until its appearance in the cuttings corresponding to that depth, in spite of:

- a very clear decrease in the drilling time,
- a significant increase in the mud salinity. Measurement and continuous recording of the mud resistivity with a resistivity (conductivity) measuring device placed in the mud return line would most probably have been a better indicator of the phenomenon than the salinity value entries on the geological log (see appendix).

The additional information given by the electrical logs, which were not available to the geologist at the time of drilling, leads to the lithological cross-section given in Figure 2. The logs have to be depth-corrected up 10 feet.

Thirteen representative levels were plotted on a Mineral Identification Plot (MID plot, Figure 3). The values from the electrical logs are raw values without any particular correction, in spite of the hole being in bad condition in places, which explains the somewhat scattered points.

The selected levels may be grouped in three families:

- Points 5, 7, 9 and 10 are representative of the saliferous beds, while point 13 is taken in the massive salt which most probably contains some impurities.

- Points 4 and 8 are taken in a type of shale encountered in this formation. Points 1, 6 and 12 are levels influenced by this type of shale.

  The largest cavings suggest that there may be an other type of shale. Reading of the electrical logs in front of these caves becomes very delicate if not impossible, therefore no points were selected for these levels.

- Points 2 and 3 give the sandstone matrix, which is encountered only at very few levels.

The only point left, point 11, is most probably a mixture of salt and shale and possibly sandstone. The plot of this point is doubtful, the neutron curve being difficult to read for this level.

The lithological cross-section may thus be summarized as follows:

The upper section of the interval shows a lower proportion of sandstone than indicated by the cutting analysis. The first salt bed appears at 8 610', i.e. when the rate of penetration sharply increases. The saliferous section is interrupted at 8 695' and replaced by an unconsolidated shaly formation with a sandstone bed in the middle of this interval.

Salt appears again at 8 770', at which depth the drill string got stuck. The electrical logs indicate no massive anhydrite bed.

The caliper curve shows a large cave starting at the top of the salt. However, the hole is more caved in front of the shale beds (caliper fully opened to 16") than in front of the salt (diameter is around 12"). Starting at 8 774', the caliper shows a very distinct improvement in hole shape due to the use of oil-base mud.

The shaly levels are well defined by the Gamma-Ray in the upper and lower section of this example where the hole, even when caved, stays at a reasonable diameter. In the much caved middle section, the Gamma-Ray is not representative of the formation.

The study of the gas-detector curve is interesting because it indicates that the two intervals with gas shows correspond to mostly shaly intervals with a few interbedded sandstone layers. The detected gas most probably comes from the shales. Indeed, there is no gas show for the saliferous layers.

5. CONCLUSION

The upper saliferous beds were drilled with a non-saturated mud and the salt did not show in the cuttings brought back to the surface. Thus, the collected cuttings were not representative of the formation; in particular, the sandstone proportion was much exaggerated.

However, the constant increase in salinity, the sharp increase in drilling speed and the total disappearance of all indication of gas are correlatable phenomena which must lead to a suspicion of the presence of salt. One may even consider as exceptional the fact that salt was detected in the cuttings when the salt content in the mud was only 90 000 ppm. This was unfortunately too late as this observation was made while circulating with the drill string stuck at 8 774'.

APPENDIX

1. In the petroleum industry, the quantity of salt contained in a solution is usually expressed

- either in grams per litre of solution,
- or in parts per million (milligrams per kilo).

The resistivity of a salt solution decreases when

- the quantity of dissolved salts increases,
- the temperature increases.

The graph of Figure 1 gives

- the conversion of grams per litre into parts per million and conversely,
- the resistivity of a solution as a function of its temperature and its estimated salinity in equivalent NaCl concentration,
- the salinity of a solution (in equivalent NaCl) knowing its resistivity and temperature.

Example 1: The resistivity of a solution containing 150 g/l of salts (equivalent NaCl) at 30°C is 0.055 ohm.metre.

Example 2: A solution with a resistivity of 0.1 ohm.metre at 80°C contains 80g/l of salts (equivalent NaCl).

2. The previously discussed monograph is adapted to solutions containing only NaCl. When the salinity is expressed in chlorine ions Cl^-, the result must be multiplied by 1.645 to obtain the salinity in equivalent NaCl.

When other salts are dissolved in the sample, correction factors must be used to take into account the ionic activity of each salt. By following the method recommended by Schlumberger, the graph of Figure 2 makes it possible to calculate, from the analysis of a solution containing various salts, the equivalent NaCl concentration of the most commonly encountered ions *. This method does not have the degree of accuracy of a resistivity measurement using the solution itself.

* Other authors or companies recommend different methods. The results obtained may be different.

Example: The analysis of a solution produced the results given in column 1 of the following table. From the graph, the multipliers for the total solid concentration are obtained (column 2).
The equivalent NaCl concentrations are given in column 3.

	Analysis (ppm)	Multiplier	Equivalent NaCl (ppm)
Na^+ Cl^-	33 000	1	33 000
SO_4^{--}	12 000	0.36	4 320
Mg^{++}	7 000	0.88	6 160
Ca^{++}	5 000	0.78	3 900
Total solid concentration	57 000		
Equivalent NaCl concentration			47 380

CASE 12 Fig. 2

MASTER LOG – WIRE LINE LOGS ASSEMBLY

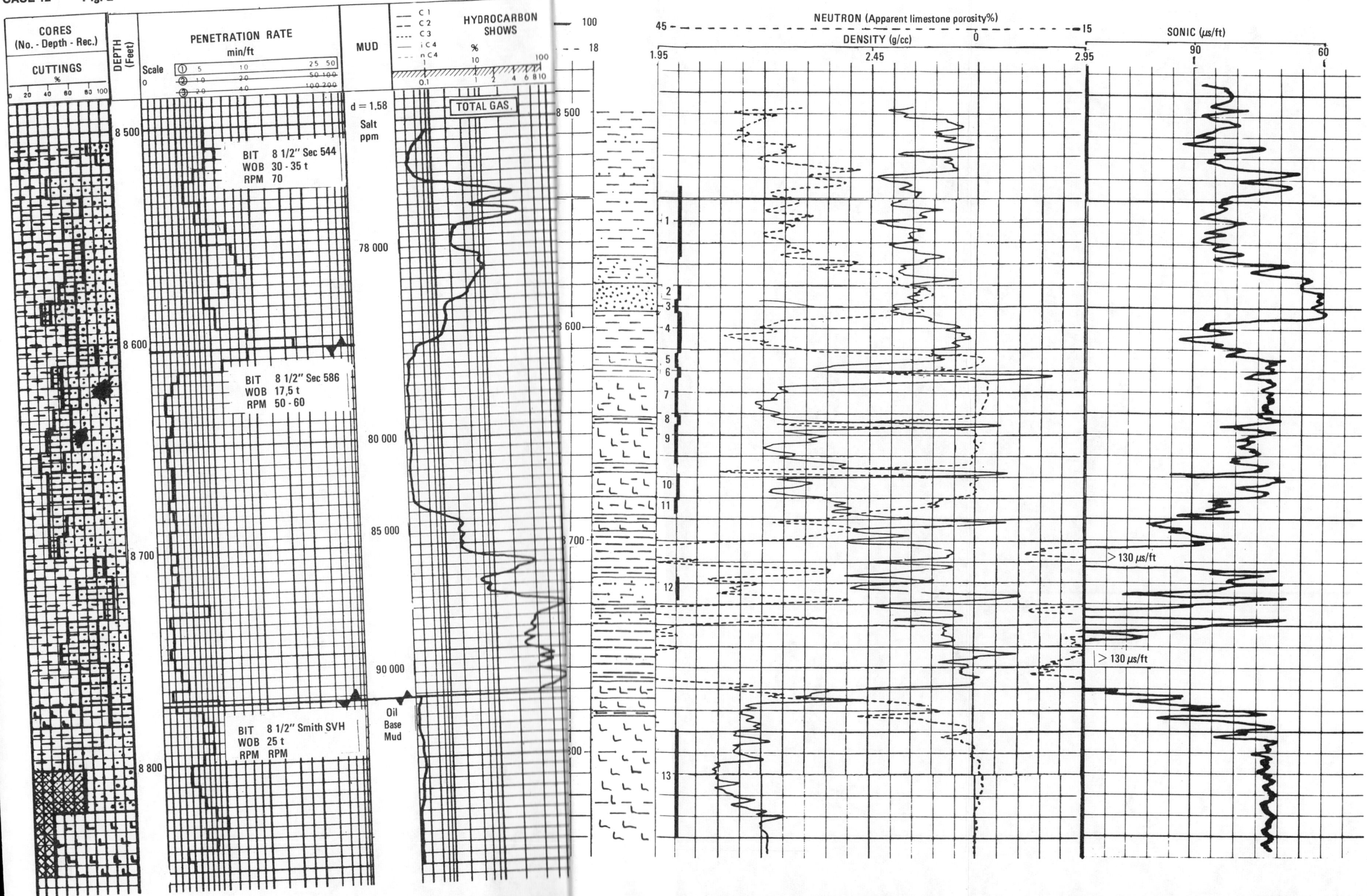

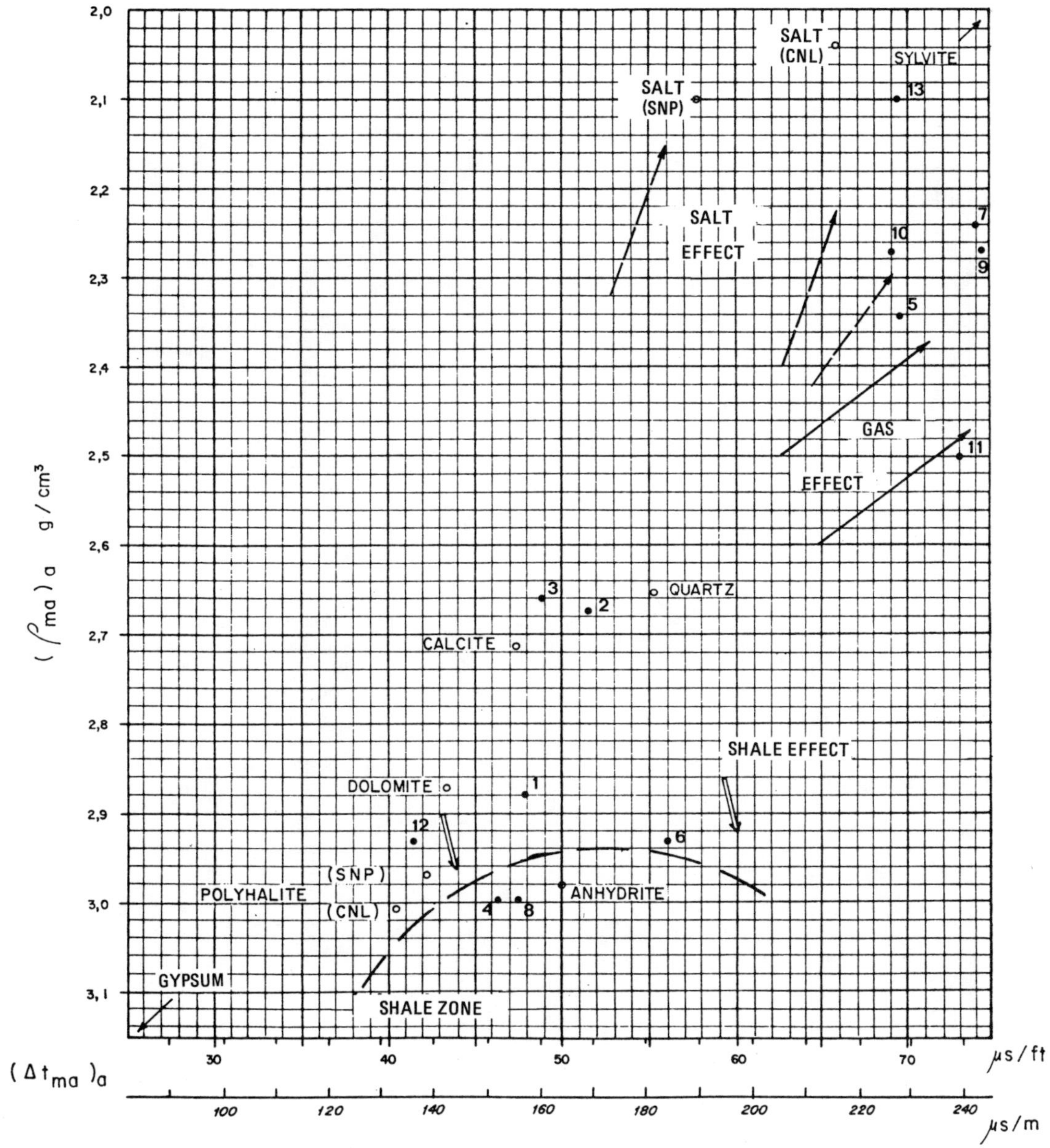

Fig. 3

MINERAL IDENTIFICATION PLOT (MID PLOT*)

*After Schlumberger - Chart CP-15 - Chart book 1978

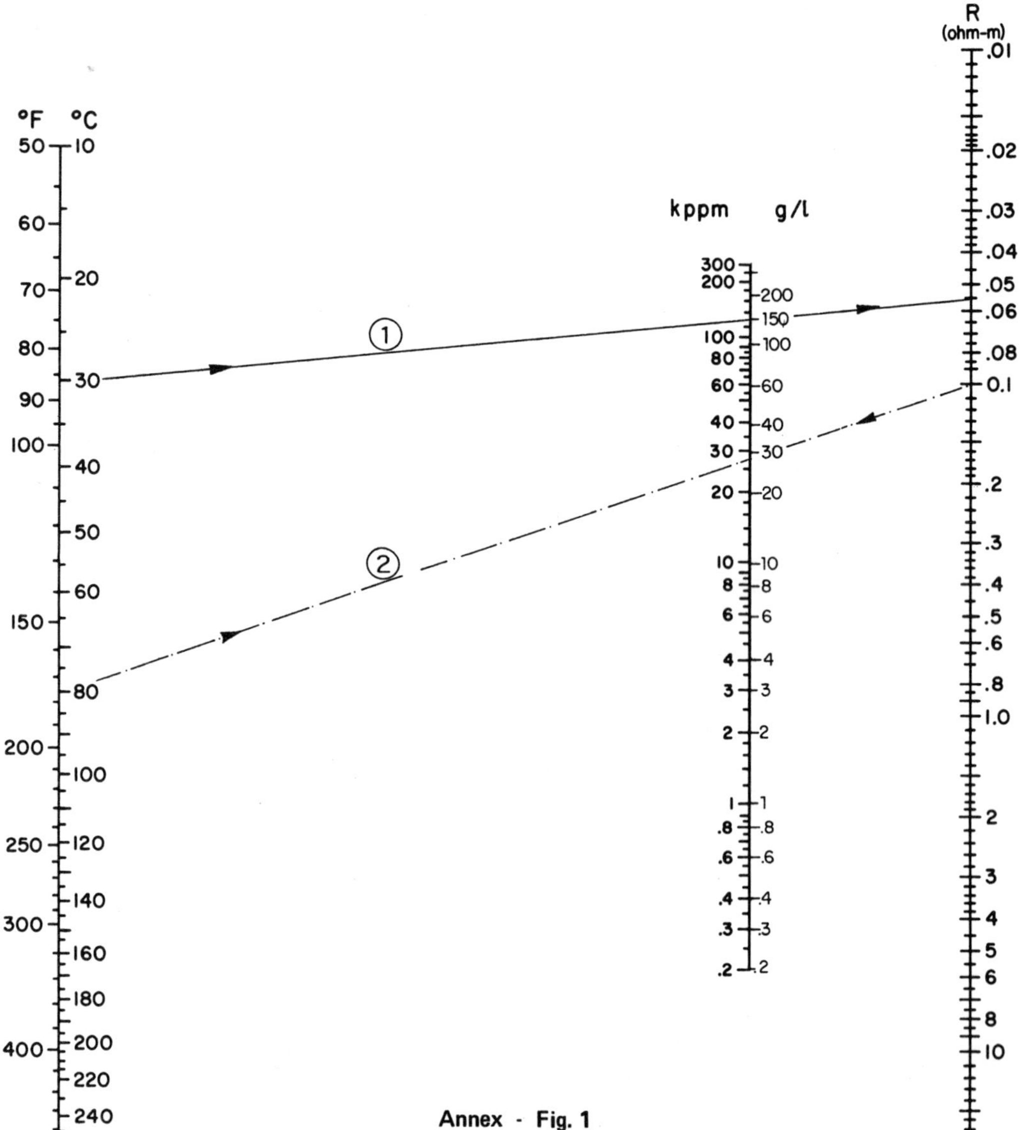

Annex - Fig. 1

RESISTIVITY NOMOGRAPH FOR NaCl SOLUTIONS

Nomograph after Schlumberger - Chart Gen-9 - Chart book 1977

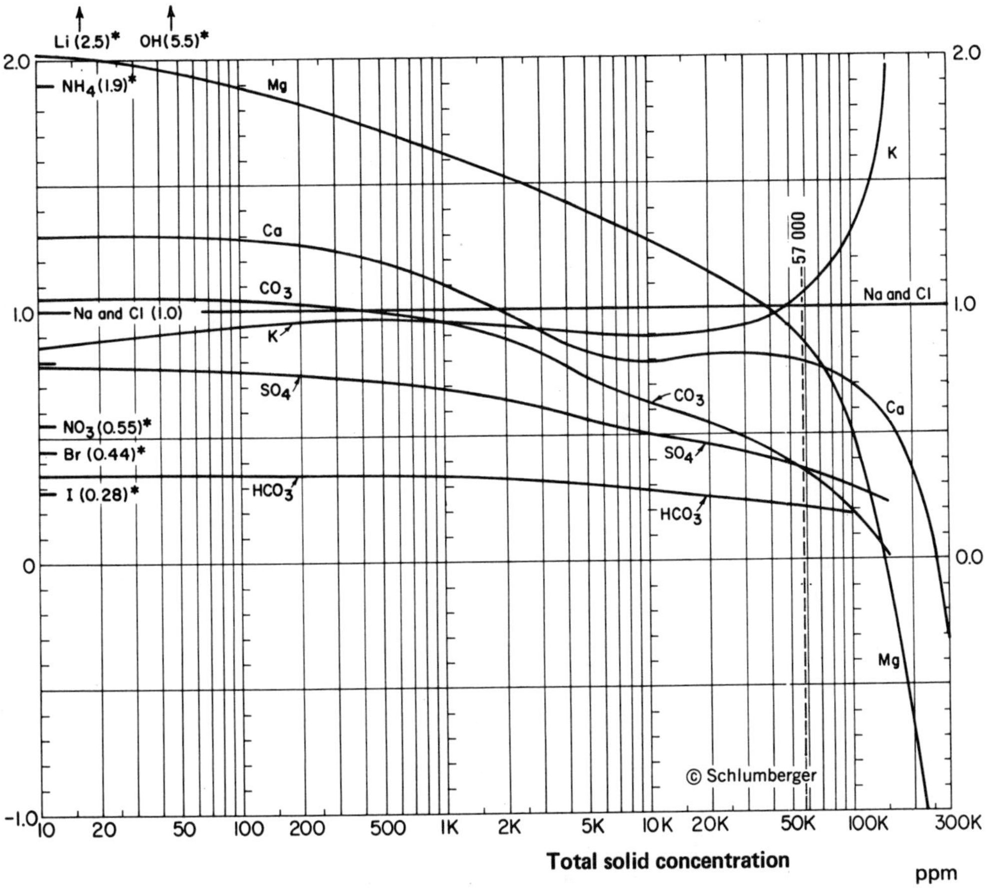

Annex - Fig. 2

EQUIVALENT NaCl CONCENTRATION DETERMINATION CHART FOR WEIGHTING MULTIPLIERS FINDING.

After Schlumberger - Chart Gen-8 - Chart book 1977

DIAMOND-BIT TURBODRILLING THROUGH AN EVAPORITIC SERIES

Using a turbine with a diamond bit may considerably modify the aspect and representativeness of the cuttings. The case studied here illustrates this sort of problem: it shows the 2 910 to 3 000-metre section of a well drilled in 1976 in an evaporitic series. This type of formation was not foreseen in the drilling programme, as it had not been encountered in previously drilled neighbouring wells.

1. TECHNICAL DATA

- Depths are in metres.
- Drilling is proceeding at 2 910 metres with a turbine fitted with a diamond tool; bit size 8 1/2".
- Since the salt was not expected, soft water-base mud (5 000 ppm) is used here. A measurement of the mud filtrate resistivity at the top of the section studied indicates a resistivity of 0.95 ohm.metre at 21°C. This corresponds to a salinity of 6 500 ppm and seems to indicate increased quantities of salts dissolved in the mud.

2. AVAILABLE DOCUMENTS

Figure 1 is the 1/500 scale log recorded at the well site. It shows the following information, from the left to the right:

- Calcimetry (or rather: carbonate content), with two curves corresponding to readings at 1 and 15 minutes.
- Cutting contents in percentages.
- Lithological cross-section drawn by the field geologist as drilling progresses.
- Depth scale.
- Rate of penetration in minutes per metre, scale 2: 0 to 100 min/m full-scale, recorded every half-metre.
- Mud data:
  - filtrate resistivity
  - mud characteristics: density, viscosity, filtrate

- Hydrocarbon shows: composition in percent of the gas-air mixture coming from the degasser installed in the flow-line close to the riser. The actual quantity of gas contained in the mud was also measured. This punctual analysis was done after complete degassing of the mud under vacuum.
- Description of the cuttings collected at the shale-shakers by the field geologist.

Figure 2 is a composite log made with parts of the geological log and electrical logs run in the open hole. It shows the left section of the mud log with the rate of penetration, the Gamma Ray-Caliper, the interpretative lithological cross-section and the Formation Density, Neutron and Sonic logs.

2. OPERATION DISCUSSION (Figure 1)

2 910 to 2 949 metres

The penetration rate is moderate over this interval: 10 to 24 min/m. However it shows some wide contrasts:

- 2 910-2 918 metres: 10 to 12 min/m
- 2 918-2 931.5 metres: 11 to 19.5 min/m
- 2 931.5-2 936.5 metres: 9 to 11 min/m (with a peak at 12.5 for 0.5 m)
- 2 936.5-2 947.5 metres: 11 to 19 min/m (with two peaks at 24 for 0.5 m)

The penetration rate increases rapidly in the last one and a half metre of this interval.

From the geologist's description, the grey gypseous shale which is predominant in the first ten metres of this example is replaced at 2 920 metres by a white gypseous shale with traces of gravel or pebbles of micritic beige limestone and traces of milky quartz. Carbonate content decreases from 30% to 10%.

At 2 940 metres, the grey gypseous shale reappears, representing about 50% of the cuttings.

At 2 920 metres, the mud filtrate resistivity which is 0.95 ohms.metre at 21°C corresponds to 6 500 ppm NaCl equivalent.

No hydrocarbon show for this interval.

2 949 to 3 000 metres

After a short transition between 2 947.5 and 2 949 metres, the penetration rate becomes notably higher: it is frequently between 4 and 5 min/m down to 2 991.5 metres, but with distinct slowdowns reaching 8 to 10 min/m. Starting at 2 991.5 metres the penetration rate stays at about 8 min/m.

At 2 952 metres, the geological circulation was motivated by the fast drilling rate recorded since 2 949 metres. Drilling is then resumed normally.

The cutting description indicates massive gypsum (with a question mark) starting at 2 947.5 metres. The logged cutting percentages indicate an increase in gypsum down to 2 965 metres where it reaches 60%. It stays at that value down to 2 984 metres. The gypsum percentage then decreases to nearly zero at 2 996 metres. The white gypseous shale has also disappeared. Beige gypseous marls appear at 2 984 metres and are predominant at the bottom of the interval. The carbonate content slowly increases to 40% at 2 990 metres and then stays at this value.

There is still no hydrocarbon show, and a measurement taken from the mud after vacuum degassing (VMS analyzer) during the geological circulation at 2 952 metres confirms the detector indication. Only an unquantified, small amount of nitrogen and hydrogen is found.

In the lower section of the interval, a mud filtrate resistivity measurement gives 0.6 ohm.metre @ 19°C, corresponding to a salinity of 11 000 ppm NaCl equivalent.

4. COMMENTS

The well-site geologist's interpretation begins to indicate gypsum and shales, with traces of coal, only at around 2 960 metres.

Even if it is difficult to foresee the presence of massive gypsum at a depth of 3 000 metres, some information of paramount importance was neglected:

- In the slow penetration rate section in the example, the lithological representation does not take into account the clear penetration rate contrasts mentioned above.

- The geologist noticed the penetration break which shows clearly between 2 947.5 and 2 949 metres, and a geological circulation was requested. However, no conclusion was drawn since this operation did not provide any new evidence that could have explained this penetration break.

- The increase in mud salinity estimated from resistivity measurements shows that the quantities of salts dissolved in the mud almost doubled within 70 metres (2 950 to 2 990 m), going from 6 500 to 11 000 ppm NaCl equivalent.

It is unfortunate that no quantitative analysis of the chlorides was performed, since it would have distinguished the salt (halite: NaCl) from the gypsum ($CaSO_4 (H_20)_2$).

The precise description (Fig. 2) which can be made using the electrical logs (which were not available to the geologist at the time of drilling), confirms the penetration-rate variations and the salinity increase. It also explains why the geological circulation did not bring any usable information.

A lithological cross-section may be drawn by correlation with the electrical log readings after positioning the evaporites' theoretical values, as was done at the bottom of Figure 2. The electrical logs have to be up-depth corrected 2.5 metres to agree with the drillers' depth.

From 2 910 to 2 949 metres, the shaly and anhydritic interbeddings, clearly characterized on the electrical logs, respectively correlate extremely well with the faster and slower drilling rates previously described in detail.

The bottom section considered in the example consists of alternating salt beds and shaly layers which also correlate very well with the sharply defined penetration-rate variations. The shale now corresponds to slow drilling rates. Notice that over the entire interval the shales were drilled at an almost constant speed of 9 to 10 min/m.

5. CONCLUSION

The cuttings' lack of validity when turbodrilling with a diamond bit may be a strong handicap for the field geologist. The fineness of the cuttings amalgamated with the drilling mud may give a very distorted representation of the drilled formation. Thus it is of paramount importance to watch closely and look among the other available sources of information for the parametres that may remedy the "invalidity" or lack of representativeness of the cuttings. In such a case, the method of differentiation between anahydrite and gypsum may be ineffective if separate characterized samples are not available (cf. recommendation "Identification of the evaporitic series while drilling").

Mud-salinity change is also a parameter well worth considering. A continuous resistivity measurement with the measuring instrument placed in the flow-line would probably have indicated a more pronounced change when entering the salt. This cannot be observed by taking occasional measurements. Also, chloride quantification should not be forgotten: in this example the NaCl would have been identified.

Last but not least, the rate of penetration is the prime parameter when drilling evaporites: all the heterogeneities later characterized using the electrical logs are found on the ROP curve. They were amost completely ignored by the field geologist except in the positioning of the shale layers at the bottom of the interval.

DRILLING THROUGH A SANDSTONE SERIES WITH SALT-CEMENTED SAND LAYERS

This example is based on the 5 300 to 5 630-foot interval of a well drilled in 1976 in a sandstone series with shale layers. These sandstones were found to be locally salt-cemented.

1. TECHNICAL DATA

- Depths are in feet.
- 13 3/8" casing shoe at 4 633'. Drilling in progress at 5 300', bit size 12 1/4".
- Drilling with saltwater-base mud; salinity increases from 65 000 ppm at 4 900' to 147 000 ppm at 5 250' i.e. 50' above the interval studied. Also, the mud density which was 1.07 at 4 900' increases progressively. It is 1.15 at 5 300'.

2. AVAILABLE DOCUMENTS

Figure 1 is the 1/500 scale log recorded at the well site. It shows the following information, from the left to the right:

- Cutting contents in percentages
- Depth scale
- Rate of penetration in minutes per foot, scale 1: 0 to 50 min/ft
- Mud data:
  . salinity in parts per million (ppm)
  . resistivity of the mud filtrate (Rmf) in ohm.metre
  . density
- Gas detector curve: composition in percent of methane equivalent of the gas-air mixture coming from the degasser installed in the flow-line close to the riser.
- Description of the cuttings prepared by the field geologist. The lithological cross-section was not drawn at the well-site.

Figure 2 is a composite log made with parts of the field log and electrical logs: Gamma-Ray, Caliper, Neutron, Formation Density and Sonic.

3. OPERATION DISCUSSION

Since drilling was resumed after setting the 13 3/8" casing at 4 632', and before entering the interval discussed hereafter, the penetration rate was very fast and there was almost no hydrocarbon show. The description of the samples indicates sandstones with some shale layers; no particular mention is made of the sandstone cement. The increase in salinity from 65 000 ppm to 147 000 ppm over 350' was noted without any comment.

Discussion of the interval presented (Fig. 1):

5 300 to 5 405 feet

Drilling is proceeding rather fast at 2 to 3 min/ft down to 5 355', then a little more slowly down to 5 405'. The slight slowdown to 4.5 ft/min seems to correspond to higher percentages of red shale and grey silt observed between 5 350 and 5 370'; these percentages then quickly decrease. Apart from these limited quantities of shales, the cuttings consist of coarse white sandstone without any conglomeratic aspect. The gas curve remains at very low values with a very fugitive peak reaching 2% between 5 355 and 5 365' and a larger peak exceeding 10% between 5 380 and 5 395'. Measurement of the mud filtrate resistivity at 5 350' indicates 0.05 ohm.metre @ 28°C, corresponding to a salinity of about 155 000 ppm.

Bit change at 5 405'.

5 405 to 5 630 feet

After drilling is resumed with a new bit, the drilling speed seems to increase: it varies between 1 and 1.5 ft/m, exceptionally reaching 2 and 2.5 ft/m. As in the previous bit run, these slight slowdowns correspond to a small increase in red shales in the cuttings. The cuttings are essentialy made of a coarse light sand, with traces of siliceous cement starting at only 5 490'. It should be noted that the significant increase in shale percentage starting at 5 600' does not result in a significant drop in penetration rate.

The gas curve stays at extremely low values, less than 0.1%, except between 5 543 and 5 570 where it peaks at 5% methane equivalent.

A mud-salinity measurement at 5 560' again shows an increase: the salt content reaches 165 000 ppm.

Density is 1.2 at 5 500'.

4. COMMENTS

The series is apparently very monotonous. There is little contrast on the rate-of-penetration curve; cuttings are almost exclusively made of sandstone and coarse sand, with two red shale and siltstone layers not exceeding 40% of the cuttings over 20'. Only at the end of the interval does the lithology start to show a more pronounced change. The three gas peaks mentioned above do not give any important information for the interpretation of the series.

The only important variation recorded is that of the mud salinity: the salt content went from 147 000 to 165 000 ppm over a 300' interval. The filtrate resistivity measured at 5 350' (0.05 ohm.metre @ 28°C) checks with the average salinity.

The log assembly shown in Figure 2 makes it possible, with the help of the electrical logs, to draw the lithological cross section shown between the caliper log and the other electrical logs.

In the upper section of the interval, the caliper shows an extremely caved hole, with a diameter larger than 17" from the top of the interval down to 5 440'. The diameter then progressively decreases to stay around 13 1/2" from 5 500' down.

The analysis of the electrical logs, using mainly the Mineral Identification Plot method (MID plot, Figure 3) with 21 typical levels as shown in Figure 2, indicates that the caved section of the hole is mostly made of salt-cemented sandstone. The points were plotted using raw values, without any particular correction in spite of the bad shape of the hole: this may have led to a certain scattering of the points. However, the salt influence is particularly noticeable on levels 1, 3, 11 to 13 and 18. Levels 4, 8, 10, 16 and 17 represent sandstone formations, possibly containing traces of carbonate although there was no mention of carbonates in the cutting description. Levels 6, 9 and 20 identify the type of shale present in this formation, moderately supported by the Gamma-ray. Last, levels 2 and 15 on the one hand, 5, 7, 14 and 19 on the other hand, show the moderate influence of salt for the first set of points, and of shale for the second set.

The quantity of saliferous cement decreases starting at 5 440', corresponding to the beginning of the better-shaped hole section. The field geologist mentioned traces of siliceous cement at 5 490'.

It is to be noted that the gas-curve peaks correspond to an increase in the interbeddings of shale and slightly- or non-saliferous sandstone. The most important peak, which reaches 11%, correlates with a clean sandstone layer, well characterized on the MID plot (point 10), and where the caliper shows a lesser caving than for the over- and underlying beds. The peak, spreading from 5 543 to 5 570', i.e. 27', corresponds to levels of the same type as those identified as sandstone with salt (18) or shale (19) influence. The sandstones found to be saliferous present most probably no characteristics of a reservoir and did not show on the gas curve.

5. CONCLUSION

The interval shown in this example represents only the lower section of a formation mostly made of sandstones that began at 4 600'. Note the lack of contrast on the rate-of-penetration curve.

From the top of this formation, the cuttings remained largely similar to those described in this example.

The increase in mud salinity is the only significant change that was noted but this information was not used. No conclusion or remark was written down in the description, and there is no mention of salt in the field geologist's report. The punctual salinity measurements are meaningful on their own. However, a continuous mud resistivity measurement at the flow-line would have better shown the presence of the saliferous cement, producing a curve, the drift of which would have been more significant.

The electrical logs permit the unambiguous identification of the salt-cemented sandstones, by combining the Neutron, Formation Density and Sonic logs. Besides, in such a case one should not rely only on the examination of the Neutron and Formation Density logs alone as this might lead to the wrong conclusion that there is gas present since neither log measures characteristic values of the salt contained in the rock.

Since drilling was conducted with a moderately salty mud, it was normal that no trace of salt was detected in the cuttings.

In such cases, one must pay great attention to the recorded data, particularly the mud salinity and gas detector indications, which will have to be taken into account when interpreting the electrical logs.